Chemical Managen
in Scientific
and Educational
Institutions

Teresa Siemieńska, Krzysztof Szczeponek,
Andrzej M. Turek, Leonard M. Proniewicz

Chemical Management in Scientific and Educational Institutions

JAGIELLONIAN UNIVERSITY PRESS

The publication was supported by Jagiellonian University – Institute of Chemistry

REVIEWER   *dr hab. Małgorzata Barańska*

COVER DESIGN   *Marcin Bruchnalski*

EDITOR   *Jerzy Hrycyk*

TECHNICAL EDITOR   *Ewelina Korostyńska*

PROOFREADER   *Katarzyna Kierejsza*

TYPESETTING   *Marek Preizner*

ISBN 978-83-233-2672-4

www.wuj.pl

Wydawnictwo Uniwersytetu Jagiellońskiego
Redakcja: ul. Michałowskiego 9/2, 31-126 Kraków
tel. (012) 631-18-80, tel./fax (012) 631-18-83
Dystrybucja: ul. Wrocławska 53, 30-011 Kraków
tel. (012) 631-01-97, tel./fax (012) 631-01-98
tel. kom. 0506-006-674, e-mail: sprzedaz@wuj.pl
Konto: PEKAO SA O/Kraków, nr 80 1240 4722 1111 0000 4856 3325

Contents

1. Introduction

Chemistry did not have such a widespread application in distant times as it has today. Alchemistry was a domain of a small group of persons and its products usually did not pose a hazard to the environment or to humans and even if this happened the extent of such hazard was limited lest to say negligible. Poor development of chemistry in later periods did not involve the adverse effects in the environment. The wastes, relatively nonhazardous to the environment and easily autodegradible, came mainly from the human activities. The present progress in civilization and especially the universal globalization involves application of more and more advanced technologies. These technologies make use of diverse human achievements including these in the field of chemistry. The rapid development of natural sciences and increasing industrial productiveness become to a larger and larger extent oppressive to the environment. Moreover, the scientific research and educational processes have become in and of themselves a source of very hazardous wastes. In consequence a serious ecological problem has arisen calling for central regulatory decisions to be taken.

Purchase, management and use of chemical reagents as well as the waste recovery and disposal are regulated by numerous legal acts. Despite the factual knowledge related to dealing with chemical reagents the knowledge of the relevant regulations within that scope is firmly required. Actually, one may seek in vain for a single study that would contain the information on a full "migration" pathway of a chemical reagent viewed in the light of legally binding regulations – from purchase, through use, storage till recovery and disposal of the derived waste. Accumulation of a full knowledge of these facts is time-consuming and sometimes even hardly accomplishable due to the complexity of the problem.

The present study is a committed to paper collection of experiences gained during many years of work in the afore mentioned domain. It is addressed to all kinds of institutions (research, academic, schools and the like) that conduct the activities involving a contact with chemicals.

A full pathway followed by a reagent from a moment of purchase, through storing, giving over, handling and using at the workplace, collecting waste, transferring to recovery and disposal has been traced taking into account the European Union and Polish legislations.

For clarity this study has been divided into chapters containing an in-debth discussion of the most important questions concerning the chemical management – legal basis (regulatory acts and directives, occupational safety and health rules, fire regulations), purchase of chemical substances and preparations, storage, handling and using of chemicals at the workplace, and finally the recovery and disposal of the chemical waste.

2. Basic Legal Regulations

Legislation within the scope of the environmental protection and waste management is a very extended field, in particular after the access of Poland to the European Union. The first act on chemical waste was entered into force in 1997 and was thoroughly amended in 2001. The acts related to this act are: "Environmental Protection Law", Act on waste, Act on packagings and packaging waste, Act on obligations of entrepreneurs within the scope of waste management, Act on waste classification, and numerous local ordinances. Not complying with the regulations results in criminal responsibility in a form of a fine or even a loss of liberty.

The most important legal acts related to the waste management according to the state by 31 December 2007 are listed below.

2.1. EC legislation adopted in Poland immediately after the access to European Union (in force since 1 May 2004)

a) Regulation (EC) No 2037/2000 of the European Parliament and of the Council of 29 June 2000 on substances that deplete the ozone layer (*Official Journal L 244, of 29.09.2000, p. 1–24, with later amendments*).

b) Regulation (EC) No 2150/2002 of the European Parliament and of the Council of 25 November 2002 on waste statistics (*O. J. L 332, of 09.12.2002, p. 1, with later amendments*).

c) Regulation (EC) No 850/2004 of the European Parliament and of the Council of 29 April 2004 on persistent organic pollutants and amending Directive 79/117/EEC (*O. J. L 158, of 30.04.2004, p. 7–49, with later amendments*).

d) Commission Regulation (EC) No 784/2005 of 24 May 2005 adopting derogations from the provisions of Regulation (EC) No 2150/2002 of the European Parliament and of the Council on waste statistics as regards Lithuania, Poland and Sweden (*O. J. L 131 of 25.05.2005, p. 42*).

e) Regulation (EC) No 166/2006 of the European Parliament and of the Council of 18 January 2006 concerning the establishment of a European Pollutant Release and Transfer Register and amending Council Directives 91/689/EEC and 96/61/EC (*O. J. L 33 of 4.02.2006, p. 1–17*); in force from 24 February 2006.

f) Regulation (EC) No 1013/2006 of the European Parliament and of the Council of 14 June 2006 on shipments of waste (*O. J. L 190, of 12.07.2006, p. 1–98*); in force since 12.07.2007.

g) Commission Regulation (EC) No 801/2007 of 6 July 2007 concerning the export for recovery of certain waste listed in Annex III or IIIA to Regulation (EC) No 1013/2006 to certain countries to which the OECD Decision on the control of transboundary movements of wastes does not apply (*O. J. L 179, of 07.07.2007, p. 6–35*).

2.2. List of Polish legally binding acts related to the waste management

Acts

a) Act on the environmental protection law of 27 April 2001 (*JL 2001 No. 62, item 627, with later amendments in JL 2002 No. 143, item 1196, JL 2002 No. 233, item 1957, JL 2003 No. 190, item 1865, JL 2004 No. 49, item 464, JL 2005 No. 113, item 954, JL 2006 No. 50, item 360, JL 2007 No. 88, item 587, JL 2008 No. 111, item 708*); entry into force on 1 October 2001.

b) Act on waste of 27 April 2001 (*JL 2001 No. 62, item 628, with later amendments in JL 2007 No. 39, item 251 and JL 2007 No. 88, item 587*); entry into force on 1 October 2001.

c) Act of 27 July 2001 on enforcing the act on the environmental protection law, the act on waste and on amending certain other acts (*JL 2001 No. 100, item 1085, JL 2002 No. 143, item 1196, JL 2003 No. 7, item 78, JL 2003 No. 190, item 1865, JL 2004 No. 49, item 464, JL 2005 No. 113, item 954, JL 2006 No. 50, item 360, JL 2006 No. 133, item 935*); entry into force on 1 October 2001.

d) Act on packagings and packaging waste of 11 May 2001 (*JL 2001 No. 63, item 638, JL 2003 No. 7, item 78, JL 2004 No. 11, item 97, JL 2004 No. 96, item 959 and JL 2005 No. 175, item 1458*); entry into force on 1 January 2002.

e) Act on the atomic law of 29 November 2000 (*JL 2001 No. 3, item 18, with later amendments in JL 2004, No. 70, item 632, JL 2004 No. 161, item 1689, JL 2007 No. 42, item 276, JL 2008 No. 93, item 583*); entry into force on 1 January 2002.

f) Act on the aquatic law of 18 July 2001 (*JL 2001 No. 115, item 1229, with later amendments in JL 2005 No. 239, item 2019, JL 2005 No. 267, item 225, JL 2006 No. 267, item 2255, JL 2006 No. 170, item 1217, JL 2006 No. 227, item 1658, JL 2007 No. 21, item 125, JL 2007 No. 64, item 427, JL 2007 No. 75, item 493, JL 2007 No. 88, item 586*); entry into force on 1 January 2002.

g) Act on chemical substances and preparations of 11 January 2002 (*JL 2001 No. 11, item 84, JL 2001 No. 100, item 1085, JL 2001 No. 123, item 1350, JL 2002 No. 125, item 1367, JL 2002 No. 135, item 1145, JL 2002 No. 142, item 1187, JL 2003 No. 189, item 1852, JL 2004 No. 96, item 959, JL 2005 No. 121, item 1263, JL 2006 No. 171, item 1225*); entry into force on 15 February 2002.

h) Act of 5 July 2002 on ratification of the Amendment to the Basel Convention on the control of transboundary movements of hazardous wastes and their disposal (*JL 2002 No. 135, item 1142*); entry into force on 12 September 2002.

i) Act of 30 August 2002 on system for conformity assessment (*JL 2002 No. 166, item 1360, JL 2004 No. 204, item 2087, JL 2005 No. 64, item 565, JL 2006 No. 170, item 1217, JL 2006 No. 235, item 1700, JL 2006, No. 249, item 1834, JL 2007 No. 21, item 124*); entry into force on 1 January 2003.

j) Act on normalization of 12 September 2002 (*JL 2002 No. 169, item 1386, JL 2004 No. 273, item 2703, JL 2005 No. 132, item 1110, JL 2006 No. 170, item 1217*); entry into force on 12 September 2002.

k) Act of 28 October 2002 on road transport of dangerous goods (*JL 2002 No. 169, item 1671, JL 2004 No. 199, item 1671, JL 2004 No. 96, item 959, JL 2004 No. 97, item 962, JL 2045 No. 173, item 1808, JL 2005 No. 90, item 757, JL 2005 No. 141, item 1184, JL 2006 No. 249, item 1834*); entry into force on 1 January 2003.

l) Act of 27 March 2003 on planning and land development (*JL 2003 No. 80, item 717, JL 2004 No. 6, item 41, JL 2004 No. 141, item 1492, JL 2005 No. 113, item 954, JL 2005 No. 130, item 1087, JL 2006 No. 45, item 319, JL 2006 No. 225, item 1635*); entry into force on 11 July 2003.

m) Act on plant protection of 18 December 2003 (*JL 2004, No. 11, item 4, JL 2004 No. 96, item 959, JL 2004 No. 173, item 1808, JL 2004 No. 273, item 2703, JL 2005 No. 163, item 1362, JL 2006 No. 92, item 639, JL 2006 No. 170, item 1217, JL 2006 No. 171, item 1225*); entry into force on 1 May 2004.

n) Act of 20 April 2004 on substances impoverishing ozone layer (*JL 2004 No. 121, item 1263, JL 2005 No. 175, item 1458, and JL 2005 No. 203, item 1683*); entry into force on 15 June 2004.

o) Act of 29 July 2005 on amending the act on waste and certain other acts (*JL 2005 No. 175, item 1458, and JL 2006 No. 63, item 441*); entry into force on 13 October 2005.

p) Act of 29 July 2005 on waste electrical and electronical equipment (*JL 2005 No. 180, item 1495*); entry into force on 21 October 2005.

r) Act of 13 April 2007 on prevention and remedying of environmental damages (*JL 2007 No. 75, item 493*); entry into force on 30 April 2007.

s) Act of 29 June 2007 on international movements of waste (*JL 2007 No. 124, item 859*) entry into force on 12 July 2007.

Regulations, Proclamations and Resolutions

Due to a large number of regulations and relevant amendments only the most important – directly related to the activities of scientific and educational units – are listed below. The regulations are directly bound to the corresponding superior acts and resolutions adopted earlier.

Regulations and proclamations issued on the basis of specific authorization contained in the act – the environmental protection law

a) Regulation of the Ministry of Environment of 6 June 2002 on admissible limit values for certain substances in air, alert limit values for certain substances in air, and margins of tolerance for admissible limit values of certain substances (*JL 2002 No. 87, item 796*) – pursuant to Art. 86 sec. 1; in force since 12 July 2003.

b) Regulation of the Ministry of Environment of 26 July 2002 on the types of installations that may cause substancial pollution of particular natural constituents or entire environment (*JL 2002 No. 122, item 1055*) – pursuant to Art. 201 sec. 2; in force since 16 August 2002.

c) Regulation of the Ministry of Environment of 5 December 2002 on reference limit values for certain substances in air (*JL 2002 No. 1, item 12*) – pursuant to Art. 222 sec. 2; in force since 1 February 2003.

d) Regulation of the Ministry of Economy, Labour and Social Policy of 23 October 2003 on requirements for the usage and movements of asbestos and the usage and remediation of installations or equipment in which asbestos was or is used (*JL 2003 No. 192, item 1876*) – pursuant to Art. 163 sec. 1; in force since 29 November 2003.

e) Regulation of the Ministry of Environment of 9 December 2003 on substances especially hazardous for environment (*JL 2003 No. 217, item 2141*) – pursuant to Art. 160 sec. 3; in force since 1 January 2004.

f) Regulation of the Ministry of Economy and Labour of 23 July 2004 on specific requirements for certain products having negative impact on the environment (*JL 2004 No. 179, item 1846*) – pursuant to Art. 169; in force since 16 August 2004.

g) Regulation of the Ministry of Environment on the types of installations that have to be reported if exploited (*JL 2004 No. 283, item 2839*) – pursuant to Art. 153 sec. 1; in force since 1 January 2005.

h) Regulation of the Ministry of Environment of 22 December 2005 on cases in which the introduction of gases or dusts to the air from an installation does not require a permission (*JL 2004 No. 283, item 2840*) – pursuant to Art. 220 sec. 2; in force since 1 January 2005.

i) Regulation of the Ministry of Environment of 23 December on requirements for monitoring the amount of emission (*JL 2004 No. 283, item 2842*) – pursuant to Art. 148; in force since 7 January 2005.

j) Regulation of the Ministry of Economy and Labour of 28 December 2004 on the products liable to the obligation of being provided with an information essential to the environmental protection (*JL 2005 No. 6, item 40*) – pursuant to Art. 167 sec. 3; in force since 1 April 2005.

k) Regulation of the Ministry of Environment of 20 December 2005 on emission standards for installations (*JL 2005 No. 260, item 2181*) – pursuant to Art. 145 sec. 1 par.. 1 and Art. 146 sec. 2 and 4; in force since 1 January 2006.

l) Regulation of the Ministry of Economy of 27 March 2007 on specific requirements for the restriction of the use of certain substances in electrical and electronic equipment *that may have negative impact on the environment* (*JL 2007 No. 60, item 457*) – *pursuant to Art. 169 sec. 1*; in force since 3 May 2007.

m) Regulation of the Ministry of Environment of 5 November 2007 on specific conditions for granting public assistance to the undertakings being investments aimed at adapting the waste disposals to the requirements of environmental protection (*JL 2007 No. 209, item 1516*) – pursuant to Art. 405 sec. 4; in force since 13 November 2007.

n) Resolution of the Lower House of the Polish Parliament (Sejm) on accepting "State Ecological Policy for 2003–2006, with a perspective for 2007–2010" (*OJ 2003 No. 33, item 433*).

Regulations and resolutions issued on the basis of specific authorization contained in the Act on waste

a) Regulation of the Ministry of Environment of 27 September 2001 on catalogue of wastes (*JL 2001 No. 112, item 1206*) – pursuant to Art. 2001 sec. 1 par. 1; in force since 1 January 2002.

b) Regulation of the Ministry of Environment of 9 November 2002 on certifying waste management qualifications (*JL 2001 No. 140, item 1584 and JL 2005 No. 42, item 411*) – pursuant to Art. 49 sec. 8; in force since 26 December 2001; repealed with effect on 1 January 2008.

c) Regulation of the Ministry of Environment of 11 December 2001 on the types of waste or waste amounts that are under no obligation of waste accounting and on the category of small and medium-sized enterprises which are allowed to conduct a simplified waste accounting (*JL 2001 No. 152, item 1735*) – pursuant to Art. 36 sec. 13; in force since 1 January 2002.

d) Regulation of the Ministry of Environment of 11 December 2001 on information scope and specimens of forms for preparing and forwarding listings of compiled data (*JL 152 No. 152, item 1737*) – pursuant to Art. 37 sec. 5; in force since 1 January 2003, repealed with effect on 1 January 2008.

e) Regulation of the Ministry of Economy of 21 March 2002 on requirements concerning thermal treatment of wastes (*JL 2002 No. 37, item 339 and JL 2005 No. 1, item 2*) – pursuant to Art. 47; in force since 27 April 2002.

f) Regulation of the Ministry of Infrastructure of 19 December 2002 on the extent and method of applying the regulations on road transport of dangerous goods to the transportation of hazardous waste (*JL 2002 No. 236, item 1986*) – pursuant to Art. 11 sec. 5; in force since 1 January 2003.

g) Regulation of the Ministry of Health of 23 December 2002 on the types of medical and veterinary wastes that are banned from recycling (*JL 2002 No. 8, item 103*) – pursuant to Art. 42 sec. 2; in force since 8 February 2003.

h) Regulation of the Ministry of Health of 23 December 2002 on admissible methods and conditions of medical and veterinary waste disposal (*JL 2003 No. 104, item 104 and JL 2004 No. 200, item 2061*) – pursuant to Art. 42 sec. 3; in force since 8 February 2003.

i) Regulation of the Ministry of Environment of 9 April 2003 on preparing waste management plans (*JL 2003 No. 66, item 620 and JL 2006 No. 46, item 333*) – pursuant to Art. 15 sec. 8; in force since 2 May 2003.

j) Regulation of the Ministry of Economy, Labour and Social Policy of 23 December 2003 on the types of waste which collection or tranportation is not liable to waste management licensing (*JL 2004 No. 16, item 154 and JL 2006 No. 136, item 965*) – pursuant to Art. 33 sec. 4; in force since 19 February 2004.

k) Regulation of the Ministry of Environment of 13 May 2004 on conditions in which the waste is recognized as non-hazardous (*JL 2004 No. 128, item 1347*) – in pursuant to Art. 4 sec. 1, par. 2; in force since 19 June 2004.

l) Regulation of the Ministy of Economy and Labour of 4 August 2004 on specific method of waste oil treatment (*JL 2004 No. 192, item 1968*) – in pursuant to Art. 7 sec. 4; in force since 18 September 2004.

m) Regulation of the Ministry of Environment of 30 August 2004 on the waste metal receipt form (*JL 2004 No. 197, item 2033*) – pursuant to Art. 43a sec. 8; in force since 24 September 2004.

n) Regulation of the Ministry of Economy and Labour of 25 October 2005 on specific treatment of waste packagings (*JL 2005 No. 219, item 1858*) – pursuant to Art. 7 sec. 4; in force since 15 November 2005.

o) Regulation of the Ministry of Environment of 14 February 2006 on specimens of documents used for needs of waste accounting (*JL 2006 No. 30, item 213*) – pursuant to Art. 36 sec. 14; in force since 11 March 2006.

p) Regulation of the Ministry of Environment of 21 March 2006 on recovery or disposal of wastes outside the installations and equipment (*JL 2006 No. 49, item 356*) – pursuant to Art. 13 sec. 3a; in force since 27 March 2006.

r) Regulation of the Ministry of Environment of 21 April 2006 on the list of classified wastes that a waste holder can pass on to a natural person or to an organizational units not being entrepreneurs, and on admissible methods of waste recovery (*JL 2006 No. 75, item 527*) – pursuant to Art. 33 sec. 3; in force since 19 May 2006.

s) Regulation of the Ministry of Environment of 25 May 2007 on information scope and specimens of forms for preparing and forwarding listings of compiled data (*JL 2007 No. 101, item 686*) pursuant to Art. 37 sec. 5; in force since 1 January 2008.

t) Regulation of the Ministry of Health of 23 August 2007 on specific method of medical waste treatment (*JL 2007 No. 162, item 1153*) – pursuant to Art. 7 sec. 4; in force since 23 September 2007.

u) Regulation of the Ministry of Environment of 20 December on certifying waste management qualifications (*JL 2007 No. 247, item 1841*) – pursuant to Art. 49 sec. 8; in force since 1 January 2008.

w) Resolution of the Council of Ministers No. 233 of 29 December 2006 on "National Plan of Waste Management 2010" (*OJ 2006 No. 90, item 946*) – pursuant to Art. 14 sec. 4; in force since 1 January 2007.

Regulations issued on the basis of specific authorization contained in the Act on packagings and packaging waste

a) Regulation of the Ministry of Environment of 30 December 2002 on the content of lead, cadmium, mercury and sixvalent chromium in packagings (*JL 2002 No. 241, item 2095 and JL 2006 No. 183, item 1362*) – pursuant to Art. 5 sec. 3; in force since 1 January 2003.

b) Regulation of the Ministry of Environment of 8 April 2003 on the method of establishing a sum of contents of lead, cadmium, mercury and sixvalent chromium in packagings (*JL 2003 No. 183, item 619*) – pursuant to Art. 5 sec. 2; in force since 2 May 2003.

c) Regulation of the Ministry of Environment of 23 April 2004 on determining the formats for labeling of packagings (*JL 2004 No. 94, item 927*) – pursuant to Art. 6 sec. 5; in force since 1 May 2004.

d) Regulation of the Ministry of Health of 24 August 2004 on the value of deposit for unit packagings of certain hazardous agents (*JL 2004 No. 202, item 2078*) – pursuant to Art. 10 sec. 5; in force since 1 October 2004.

e) Regulation of the Ministry of Environment of 31 December 2004 on the voivodeship reports regarding the packaging management (*JL 2005 No. 4, item 29*) – pursuant to Art. 19 sec. 3; in force since 25 January 2005.

f) Regulation of the Ministry of Environment of 31 December 2004 on the formats of questionnaires for annual reporting on weight of the manufactured, brought from abroad and exported packagings (*JL 2005 No. 4, item 30*) – pursuant to Art. 7 sec. 2 and Art. 9 sec. 4; in force since 25 January 2005.

Regulation issued on the basis of specific authorizations contained in the Act on dangerous substances and preparations

a) Regulation of the Ministry of Health of 14 March 2003 on the method of labeling of the sites, pipelines as well as containers and tanks for storing of or containing hazardous substances or preparations (*JL 2003 No. 61, item 552*) – pursuant to Art. 27 sec. 2; in force since 11 July 2003.

b) Regulation of the Ministry of Health of 4 June 2003 on criteria that should be met by organizational units performing research on chemical substances or preparations and on control of meeting these criteria (*JL 2003 No. 116, item 1103 and JL 2005 No. 4, item 31*) – pursuant to Art. 24 sec. 2 par. 2–5 and sec. 4; in force since 19 July 2003.

c) Regulation of the Ministry of Health of 2 September 2003 on criteria and method of classification of chemical substances and preparations (*JL 2003 No. 171, item 1666, JL 2004 No. 243, item 2440 and JL 2007 No. 174, item 1222*) – pursuant to Art. 4 sec. 2; in force since 17 October 2003.

d) Regulation of the Ministry of Health of 2 September 2003 on labeling of packagings of dangerous substances and preparations (*JL 2003 No. 173, item 1679 and JL 2004 No. 260, item 2595*) – pursuant to Art. 26; in force since 21 October 2003.

e) Regulation of the Ministry of Health of 28 July 2003 on the methods of conducting research of the physico-chemical properties, toxicity and ecotoxicity of chemical substances and preparations (*JL 2003 No. 232, item 2343 and JL 2005 No. 251, item 2119*) – pursuant to Art. 24 sec. 3 par. 1; in force since 15 January 2004.

f) Regulation of the Ministry of Health of 28 September 2005 on the list of dangerous substances including the applied classification and labeling (*JL 2005 No. 201, item 1674*) – pursuant to Art. 4 sec. 3; in force since 29 October 2005.

g) Regulation of the Ministry of Health of 13 November 2007 on chemical safety card (*JL 2007 No. 215, item 1588*) – pursuant to Art. 5 sec. 5; in force since 16 November 2007.

Regulations issued on the basis of specific authorization contained in the Act on prohibition on using the products containing asbestos

a) Regulation of the Ministry of Economy, Labour and Social Policy of 2 April 2004 on the methods and conditions of safety usage and removal of the products containing asbestos (*JL 2004 No. 71, item 649*) – pursuant to Art. 4 sec. 1; in force since 6 May 2004.

b) Regulation of the Ministry of Economy and Labour of 14 October 2005 on the occupational safety and health rules applying to the secure isolation and removal of products containing asbestos and on the training programme on safety usage of such products (*JL 2005 No. 216, item 1824*) – pursuant to Art. 4 sec. 2; in force since 15 November 2005.

Regulations issued on the basis of specific authorization in the Act on substances impoverishing the ozone layer

a) Regulation of the Ministry of Economy and Labour of 6 August 2004 on format of the card for equipment and installation containing the controlled substances (*JL 2004 No. 184, item 1903*) – pursuant to Art. 7 sec. 3; in force since 9 September 2004.

b) Regulation of the Ministry of Economy and Labour of 11 August 2004 on accounting of the controlled substances (*JL 2004 No. 185, item 1911*) – pursuant to Art. 5 sec. 3; in force since 10 September 2004.

c) Regulation of the Ministry of Economy and Labour of 16 August 2004 on the method of labeling of the products, equipment and installations containing the controlled substances as well as the containers containing such substances (*JL 2004 No. 195, item 2007*) – pursuant to Art. 6 sec. 3; in force since 22 September 2004.

d) Regulation of the Ministry of Economy and Labour of 16 August 2004 on control of tightness of the equipment and installations containing the controlled substances (*JL 2004 No. 195, item 2008*) – pursuant to Art. 8 sec. 5; in force since 22 September 2004.

e) Regulation of the Ministry of Economy and Labour of 15 August 2004 on programmes of courses, conducting examinations and format of a qualification certificate for work with the controlled substances (*JL 2004 No. 195, item 2009*) – pursuant to Art. 12 sec. 4; in force since 22 September 2004.

f) Regulation of the Ministry of Economy and Labour of 2 September 2004 on specific requirements for technical equipment used while performing activities with the controlled substances (*JL 2004 No. 202, item 2071*) – pursuant to Art. 9 sec. 5; in force since 1 October 2004, pursuant to Art. 19 sec. 3; in force since 5 February 2005.

Regulations issued on the basis of specific authorizations contained in the Act on electrical and electronical equipment

a) Regulation of the Ministry of Environment of 29 November 2005 on format of a list of facilities processing the disposed equipment and on method of forwarding this format (*JL 2005 No. 241, item 2036*) – pursuant to Art. 33 sec. 5; in force since 1 October 2006.

b) Regulation of the Ministry of Environment of 3 January 2006 on format of reports on waste from the disposed equipment and on method of forwarding this format (*JL 2006 No. 5, item 34*) – pursuant to Art. 56 sec. 2; in force since 1 July 2006.

c) Regulation of the Ministry of Environment of 5 January 2006 on format of a report on amount and weight of the brought in equipment and on method of forwarding this format (*JL 2006 No. 6, item 37*) – pursuant to Art. 24 sec. 3; in force since 1 July 2006.

d) Regulation of the Ministry of Environment of 9 January 2006 on method of laying down the register number (*JL 2006 No. 6, item 39*) – pursuant to Art. 10 sec. 3; in force since 1 July 2006.

e) Regulation of the Ministry of Environment of 9 January 2006 on formats of reports on disposed equipment including method of forwarding these formats (*JL 2006 No. 6, item 40*) – pursuant to Art. 31 sec. 2; in force since 1 July 2006.

f) Regulation of the Ministry of Environment of 20 January 2006 on format of a certificate on disposed equipment and on method of forwarding this format (*JL 2006 No. 21, item 160*) – pursuant to Art. 50 sec. 8; in force since 24 February 2006.

g) Regulation of the Ministry of Environment of 30 January 2006 on format of a certificate on recycling and format of a certificate on a recovery treatment other than recycling (*JL 2006 No. 27, item 203*) – pursuant to Art. 55; in force since 8 March 2006.

2.3. List of legally binding EC acts related to the waste management

A. General requirements

a) Directive 2006/12/EC of the European Parliament and of the Council of 5 April 2006 on waste (*OJ L 114, 27.04.2006, p. 9–21*); in force from 17 May 2006.

b) 94/741/EC: Commission Decision of 24 October 1994 concerning questionnaires for Member States reports on the implementation of certain Directives in the waste sector (implementation of Council Directive 91/692/EEC) (*OJ L 296, 17.11.1994, p. 42–55*).

c) Commission Decision of 3 May 2000 replacing Decision 94/3/EC establishing a list of wastes pursuant to Article 1(a) of Council Directive 75/442/EEC on waste and Council Decision 94/904/EC establishing a list of hazardous waste pursuant to Article 1(4) of Council Directive 91/689/EEC on hazardous waste (*OJ L 226, 06.09.2006, p. 3–24, with later amendments*).

d) Council Directive 96/61/EC of 24 September 1996 concerning integrated pollution prevention and control (*OJ L 257, 10.10.1996, p. 26–40, with later amendments*).

e) 76/431/EEC: Commission Decision of 21 April 1976 setting up a Committee on Waste Management (*OJ L 115, 1.5.1976, p. 73–74*).

f) 81/972/EEC: Council Recommendation of 3 December 1981 concerning the re-use of waste paper and the use of recycled paper (*Official Journal L 355, 10.12.1981, p. 56–57*).

g) Council Resolution of 7 May 1990 on waste policy (*OJ C 122, 18.05.1990, p. 2–4*).

h) Council Resolution of 24 February 1997 on a Community strategy for waste management (*OJ C 76, 11.3.1997, p. 1–4*).

i) Resolution of the ECSC Consultative Committee on the classification of scrap (adopted unanimously with two abstentions during the 337th session of 10 October 1997) (*Official Journal C 356, 22.11.1997, p. 8–10*).

j) Regulation (EC) No 761/2001 of the European parliament and of the council of 19 March 2001 allowing voluntary participation by organisations in a Community eco-management and audit scheme (EMAS) (*OJ L 114, 24.04.2001, p. 1, with later amendments*).

Waste statistics, data bases

a) Regulation (EC) No 2150/2002 of the European Parliament and of the Council of 25 November 2002 on waste statistics (Text with EEA relevance) (*OJ L 332, 9.12.2002, p. 1–36, with later amendments*).

b) Commission Regulation (EC) No 784/2005 of 24 May 2005 adopting derogations from the provisions of Regulation (EC) No 2150/2002 of the European Parliament and of the Council on waste statistics as regards Lithuania, Poland and Sweden (*OJ L 131, 25.5.2005, p. 42–42*).

c) Commission Regulation (EC) No 1445/2005 of 5 September 2005 defining the proper quality evaluation criteria and the contents of the quality reports for waste statistics for the purposes of Regulation (EC) No 2150/2002 of the European Parliament and of the Council (*OJ L 229, 6.9.2005, p. 6–12*).

d) Regulation (EC) No 166/2006 of the European Parliament and of the Council of 18 January 2006 concerning the establishment of a European Pollutant Release and Transfer Register and amending Council Directives 91/689/EEC and 96/61/EC (*OJ L 33, 4.2.2006, p. 1–17*).

International waste turnover

a) The Basel Convention on the control of transboundary movements of hazardous wastes and their disposal (adopted in Basel, Switzerland on 22 March 1989 and entered into force on 5 May 1992).

b) Council Decision of 22 September 1997 on the approval, on behalf of the Community, of the amendment to the Convention on the control of transboundary movements of hazardous wastes and their disposal (Basel Convention), as laid down in Decision III/1 of the Conference of the Parties (*Official Journal L 272, 04.10.1997, p. 0045–0046*).

c) Council Resolution of 21 December 1988 concerning transfrontier movements of hazardous waste to third countries (*Official Journal C 009, 12.01.1989, p. 0001–0001*).

d) Regulation (EC) No 1013/2006 of the European Parliament and of the Council of 14 June 2006 on shipments of waste, in force from 12 July 2007 (*OJ L 190, 12.7.2006, p. 1–98*).

e) Commission Regulation (EC) No 801/2007 of 6 July 2007 concerning the export for recovery of certain waste listed in Annex III or IIIA to Regulation (EC) No 1013/2006 to certain countries to which the OECD Decision on the control of transboundary movements of wastes does not apply (*OJ L 179, 7.7.2007, p. 6–35*).

Reporting

a) Council Directive 91/692/EEC of 23 December 1991 standardizing and rationalizing reports on the implementation of certain Directives relating to the environment (*OJ L 377, 31.12.1991, p. 48–54, with later amendments*).

b) 94/741/EC: Commission Decision of 24 October 1994 concerning questionnaires for Member States reports on the implementation of certain Directives in the waste sector (implementation of Council Directive 91/692/EEC) (*OJ L 296, 17.11.1994, p. 42–55*).

c) 95/337/EC: Commission Decision of 25 July 1995 amending Decision 92/446/EEC of 27 July 1992 concerning questionnaires relating to directives in the water sector (*OJ L 200, 24.8.1995, p. 1–34*).

d) 97/622/EC: Commission Decision of 27 May 1997 concerning questionnaires for Member States reports on the implementation of certain Directives in the waste sector (implementation of Council Directive 91/692/EEC) (*OJ L 256, 19.9.1997, p. 13–19, with later amendments*).

e) 1999/412/EC: Commission Decision of 3 June 1999 concerning a questionnaire for the reporting obligation of Member States pursuant to Article 41(2) of Council Regulation (EEC) No 259/93 (notified under document number C(1999) 1456) (*OJ L 156, 23.6.1999, p. 37–46*).

f) 2000/738/EC: Commission Decision of 17 November 2000 concerning a questionnaire for Member States reports on the implementation of Directive 1999/31/EC on the landfill of waste (notified under document number C(2000) 3318) (*OJ L 298, 25.11.2000, p. 24–26*).

g) 2001/753/EC: Commission Decision of 17 October 2001 concerning a questionnaire for Member States reports on the implementation of Directive 2000/53/EC of the European Parliament and of the Council on end-of-life vehicles (notified under document number C(2001) 3096) (*OJ L 282, 26.10.2001, p. 77–80*).

h) 2004/249/EC: Commission Decision of 11 March 2004 concerning a questionnaire for Member States reports on the implementation of Directive 2002/96/EC of the European Parliament and of the Council on waste electrical and electronic equipment (WEEE) (*OJ L 78, 16.3.2004, p. 56–59*).

i) 2006/329/EC: Commission Decision of 20 February 2006 laying down a questionnaire to be used for reporting on the implementation of Directive 2000/76/EC on the incineration of waste (notified under document number C(2006) 438) (*OJ L 121, 6.5.2006, p. 38–42*).

B. Special requirements for particular methods of waste management

Waste incineration

a) Directive 2000/76/EC of the European Parliament and of the Council of 4 December 2000 on the incineration of waste (*OJ L 332, 28.12.2000, p. 91–111*).

b) Commission Decision of 20 February 2006 laying down a questionnaire to be used for reporting on the implementation of Directive 2000/76/EC on the incineration of waste (*OJ L 121, 6.05.2006, p. 38–42*).

Landfill waste

a) Council Directive 1999/31/EC of 26 April 1999 on the landfill of waste (*OJ L 182, 16.07.1999, p. 1–19, with later amendments*).

b) Council Decision of 19 December 2002 establishing criteria and procedures for the acceptance of waste at landfills pursuant to Article 16 of and Annex II to Directive 1999/31/EC (*OJ L 011, 16.01.2003, p. 27–49*).

c) Commission Decision of 17 November 2000 concerning a questionnaire for Member States reports on the implementation of Directive 1999/31/EC on the landfill of waste (*OJ L 298, 25.11.2003, p. 24–26*).

C. Special requirements for particular streams of waste

Hazardous waste

a) Council Directive 91/689/EEC of 12 December 1991 on hazardous waste (*OJ L 377, 31.12.1991, p. 20–27, with later amendments*).

b) 96/302/EC: Commission Decision of 17 April 1996 establishing a format in which information is to be provided pursuant to Article 8 (3) of Council Directive 91/689/EEC on hazardous waste (*OJ L 116, 11.5.1996, p. 26–27*).

c) 97/622/EC: Commission Decision of 27 May 1997 concerning questionnaires for Member States reports on the implementation of certain Directives in the waste sector (implementation of Council Directive 91/692/EEC) (*OJ L 256, 19.9.1997, p. 13–19*).

d) 2002/909/EC: Commission Decision of 13 November 2002 on Italian rules waiving permitting requirements for undertakings and establishments recovering hazardous waste under Article 3 of Directive 91/689/EEC on hazardous waste (*Official Journal L 315, 19.11.2002, p. 16–20*).

Animal waste

a) Regulation (EC) No 1774/2002 of the European Parliament and of the Council of 3 October 2002 laying down health rules concerning animal by-products not intended for human consumption (*OJ L 273, 10.10.2002, p. 1, with later amendments*).

Packaging waste

a) European Parliament and Council Directive 94/62/EC of 20 December 1994 on packaging and packaging waste (*OJ L 365, 31.12.1994, p. 10–23; OJ L 47, 18.2.2004, p. 26–32; OJ L 70, 16.3.2005, p. 17–18*).

b) 97/129/EC: Commission Decision of 28 January 1997 establishing the identification system for packaging materials pursuant to European Parliament and Council Directive 94/62/EC on packaging and packaging waste (*OJ L 50, 20.2.1997, p. 28–31*).

c) 97/622/EC: Commission Decision of 27 May 1997 concerning questionnaires for Member States reports on the implementation of certain Directives in the waste sector (implementation of Council Directive 91/692/EEC) (*OJ L 256, 19.9.1997, p. 13–19*).

d) 1999/177/EC: Commission Decision of 8 February 1999 establishing the conditions for a derogation for plastic crates and plastic pallets in relation to the heavy metal concentration levels established in Directive 94/62/EC on packaging and packaging waste (notified under document number C(1999) 246) (*OJ L 56, 4.3.1999, p. 47–48*).

e) 2001/171/EC: Commission Decision of 19 February 2001 establishing the conditions for a derogation for glass packaging in relation to the heavy metal concentration levels established in Directive 94/62/EC on packaging and packaging waste (*OJ L 62, 2.3.2001, p. 20–21*).

f) 2001/524/EC: Commission Decision of 28 June 2001 relating to the publication of references for standards EN 13428:2000, EN 13429:2000, EN 13430:2000, EN 13431:2000 and EN 13432:2000 in the Official Journal of the European Communities in connection with Directive 94/62/EC on packaging and packaging waste (*OJ L 190, 12.7.2001, p. 21–23*).

g) 2005/270/EC: Commission Decision of 22 March 2005 establishing the formats relating to the database system pursuant to Directive 94/62/EC of the European Parliament and of the Council on packaging and packaging waste (*OJ L 86, 5.4.2005, p. 6–12*).

h) 2006/340/EC: Commission Decision of 8 May 2006 amending Decision 2001/171/EC of the European Parliament and of the Council for the purpose of prolonging the validity of the conditions for a derogation for glass packaging in relation to the heavy metal concentration levels established in Directive 94/62/EC (*OJ L 125, 12.5.2006, p. 43–43*).

Batteries and acumulators

a) Council Directive 91/157/EEC of 18 March 1991 on batteries and accumulators containing certain dangerous substances (*OJ L 78, 26.3.1991, p. 38–41*) *– is to be repealed with effect on 26 September 2008.*

b) Commission Directive 93/86/EEC of 4 October 1993 adapting to technical progress Council Directive 91/157/EEC on batteries and accumulators containing certain dangerous substances (*OJ L 264, 23.10.1993, p. 51–52*) *– is to be repealed with effect on 26 September 2008.*

c) Directive 2006/66/EC of the European Parliament and of the Council of 6 September 2006 on batteries and accumulators and waste batteries and accumulators and repealing Directive 91/157/EEC (*OJ L 266, 26.9.2006, p. 1–14*).

Electrical and electronic waste

a) Directive 2002/95/EC of the European Parliament and of the Council of 27 January 2003 on the restriction of the use of certain hazardous substances in electrical and electronic equipment (*OJ L 37, 13.2.2003, p. 19–23*).

b) Directive 2002/96/EC of the European Parliament and of the Council of 27 January 2003 on waste electrical and electronic equipment (WEEE) – Joint declaration of the European Parliament, the Council and the Commission relating to Article 9 (*OJ L 37, 13.2.2003, p. 24–39 and OJ L 345, 31.12.2003, p. 106–107; also in Polish version: EU Official Journal – Polish special edition, chapter 15, Vol. 7, p. 359*).

c) 2004/249/EC: Commission Decision of 11 March 2004 concerning a questionnaire for Member States reports on the implementation of Directive 2002/96/EC of the European Parliament and of the Council on waste electrical and electronic equipment (WEEE) (*OJ L 78, 16.3.2004, p. 56–59*).

d) 2004/486/EC: Council Decision of 26 April 2004 granting Cyprus, Malta and Poland certain temporary derogations from Directive 2002/96/EC on waste electrical and electronic equipment (*OJ L 162, 30.4.2004, p. 114–115*).

e) 2005/369/EC: Commission Decision of 3 May 2005 laying down rules for monitoring compliance of Member States and establishing data formats for the purposes of Directive 2002/96/EC of the European Parliament and of the Council on waste electrical and electronic equipment (*OJ L 119, 11.5.2005, p. 13–16*).

Substances depleting the ozone layer

a) Regulation (EC) No 2037/2000 of the European Parliament and of the Council of 29 June 2000 on substances that deplete the ozone layer (*OJ L 244, 29.9.2000, p. 1–24; with later amendments*).

b) Regulation (EC) No 2038/2000 of the European Parliament and of the Council of 28 September 2000 amending Regulation (EC) No 2037/2000 on substances that deplete the ozone layer, as regards metered dose inhalers and medical drug pumps (*OJ L 244, 29.9.2000, p. 25–25*).

c) Regulation (EC) No 2039/2000 of the European Parliament and of the Council of 28 September 2000 amending Regulation (EC) No 2037/2000 on substances that deplete the ozone layer, as regards the base year for the allocation of quotas of hydrochlorofluorocarbons (*OJ L 244, 29.9.2000, p. 26–26*).

Asbestos

a) Council Directive 87/217/EEC of 19 March 1987 on the prevention and reduction of environmental pollution by asbestos (*OJ L 85, 28.3.1987, p. 40–45*).

Persistent organic pollution

a) Regulation (EC) No 850/2004 of the European Parliament and of the Council of 29 April 2004 on persistent organic pollutants and amending Directive 79/117/EEC (*OJ L 158, 30.4.2004, p. 7–49, with later amendments*).

b) 2007/639/EC: Commission Decision of 2 October 2007 establishing a common format for the submission of data and information pursuant to Regulation (EC) No 850/2004 of the European Parliament and of the Council concerning persistent organic pollutants (*OJ L 258, 4.10.2007, p. 39–43*).

D. Programmes and strategies

a) Decision No 1600/2002/EC of the European Parliament and of the Council of 22 July 2002 laying down the Sixth Community Environment Action Programme (*OJ L 242, 10.9.2002, p. 1–15*).

b) Communication from the Commission to the Council, the European Parliament, the Economic and Social Committee and the Committee of the Regions – Towards a Thematic Strategy for Soil Protection – COM final 2002 179 (16 April 2002).

c) Communication from the Commission to the Council, the European Parliament and the Economic and Social Committee – Towards a Thematic Strategy on the Sustainable Use of Pesticides – COM final 2002 349 (1 July 2002).

d) Communication from the Commission – Towards a thematic strategy on the prevention and recycling of waste – COM final 2003 301 (27 May 2003).

e) Communication from the Commission to the Council and the European Parliament – Towards a Thematic Strategy on the Sustainable Use of Natural Resources – COM final 2003 572 (1 October 2003).

f) Communication from the Commission to the Council and the European Parliament – A Thematic Strategy on Air Pollution – COM final 2005 446 (21 September 2005).

g) Communication from the Commission to the Council, the European Parliament, the European Economic and Social Committee and the Committee of the Regions – Taking sustainable use of resources forward – A Thematic Strategy on the Prevention and Recycling of Waste – COM final 2005 666 (21 December 2005).

3. Purchase and Storage of Chemical Reagents

Every scientific and educational institution with activity profile connected with a use of chemical reagents has to aquire the chemicals, except these not numerous synthetized on own premises, from external sources.

3.1. Purchase of chemical reagents

The best solution is a choice of few chemical supply companies out of which the two should be main. The first should supply the basic reagents for educational purposes and elementary syntheses (solvents, simple salts and substrates) – usually a national entrepreneur. The second supplier should posses a worldwide distribution network and should guarantee the possibility of supplying the specialty chemicals of high quality. A contract should be signed with these two companies clearly laying down the conditions of purchase and transportation (a possibility of ensuring through negotiations the free transport is suggested), the forms and terms of payment, the way of placing orders (i.e. by fax, e-mail, phone, etc.) and the available discount programmes. Making use of services of other suppliers is occasional – mostly due to specific characteristics of the preparations which are manufactured or supplied most often only by these companies.

Every company regardless of the scope of supplies is obliged to have an agreement of sale that is valid in a given country. Each reagent must be provided with: a quality certificate (indicating the date of applicability, the results of purity analysis, the producer and the supplier), chemical safety card (unless the relevant regulations release from this obligation). The chemical safety card includes a collection of data that are crucial for user's safety, describing in particular the hazards created by a given substance or preparation, the safety principles as regards handling and storing, and the safety measures taken in emergency situations. The above mentioned documents are oftentimes forwarded by e-mail. These documents should be then passed on a final user at the workplace or if it is impossible the final user should be instructed on another prompt, not ambarassing way of access to such information. An example of the chemical safety card is included in APPENDICES as Appendix 1.

3.2. Storage

Chemical reagents purchased in excess of the required amount (such approach is often-times justified on economic grounds) must be stored in the central storehouse. Storing chemicals at the workplace is allowed but only and solely of such amounts that are sufficient to ensure continuity of the work. The structural organization of the chemical storehouse is adjusted by relevant regulations – in particular the occupational safety and health rules and the fire regulations – and by the premises regarding the reciprocal interactions of chemicals.

3.3. Choice of storage space and structural organization of the central chemical storehouse

The ideal solution would be to mark off a separate building for a central storehouse and if it is impossible then to store the chemicals one needs to section off a part of the basement effectively isolated from other rooms and unaccessible to the outsiders. On account of the fire safety the used installations must meet the technical specifications defined by the Polish Norms and all other applicable regulations. The technical installa-tions recognized as indispensable include: the internal system of electric power supply with possibility of the power shutdown from the outside (from a control room), the ventilation system with filters installed at the vents (preferably with an air-conditionig allowing to maintain a constant temperature in the store), a source of water (taking into account a free access to it in the circumstances of a sudden fire). The used installations will require technical inspections and maintenance. In addition, the attention should be paid to as close as possible localization of a hydrant and a drive for the firemen. All doors in the store must be opened "to the outside". Each store room has to be equipped with at least two fire extinguishing media (water, sand, fire extinguishing blankets, applicable fire extinguishers – mostly dry powder fire extinguishers, and the like) and a ventilation cutoff valve located outside. The first aid medicine chest with a standard equipment must be set up in the main room in the place which is easy of approach.

Due to specific character of the stored chemicals the store should be equipped with appropriate racks, benches, shelves, and cabinets, and the floor surface in the store should be flat and easily washable.

The type of reagent to be stored on a given shelf decides on the material from which the shelf will be made or finished. Acids and dangerous substances are stored on woo-den shelves covered with an acid-proof terracotta. Toxic and extremely dangerous sub-stances should be stored in special safety cabinets. The cabinets must be locked and provided with individual ventilation connected to the central ventilation system.

Bulk quantities of chemicals must be stored in separate storage areas segregated into classes. Categorical segregation of chemicals may be based on incompatibility of the chemicals or on the hazard classes. The chemical segregation in the storehouse required as a minimum calls for:
• room for solid organic chemicals,

- room for liquid organic chemicals,
- room for inorganic chemicals,
- room for inorganic acids,
- room for dangerous substances and preparations,
- room for toxic substances with a double-locked door,
- room for storing the waste destined for recovery and disposal,
- room for radioactive substances (its construction and equipment are subject to specific regulations),
- utility room for storing empty waste containers, wheel-barrows for carrying sand destined for fire extinguishing, hand-carts for transporting chemicals, and the like.

In the respective storage rooms the rules for proper arrangement of mutually antagonistic chemicals must be preserved. This is aimed at avoiding the possibility of ensuing of unknown chemical reactions between released vapours of certain chemicals.

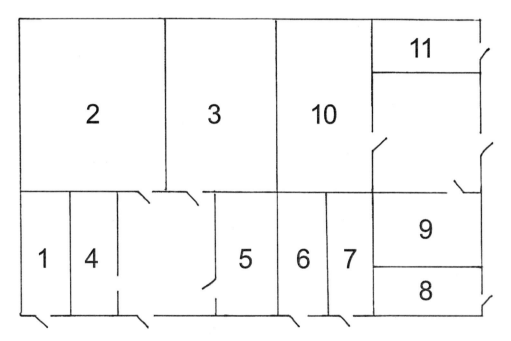

Figure 1. Scheme of the central chemical storehouse: 1, 8 – utility rooms; 2 – room for solid organic chemicals; 3 – room for inorganic chemicals; 4 – sanitary room; 5 – office room with access to the internet; 6 – room for dangerous substances and preparations; 7 – room for inorganic acids; 9 – room for toxic substances; 10 – room for liquid organic chemicals; 11 – room for wastes destined for recovery and disposal.

The store must be attended by a limited number of appropriately trained persons but large enough to ensure the continuity of operations yet without contravening the occupational safety and health rules. Inside the store during these works at least two persons must always be present. Storemans are responsible for physical conditions in the store, the actual state of chemicals, receiving and giving over the chemicals and their segregation. The store personnel must be equipped with appropriate means of personal protection including in particular the protective aprons, gloves, glasses and boots. In a storage room cotaining acids one has to wear additional protective clothing such as a rubber apron, a helmet with a face shield and acid-proof rubber gloves. Persons attending the storehouse are allowed to stay inside the store for no more than 30 minutes at a time.

In order to facilitate inventorying of chemicals and checking on the amount of chemicals in stock the storing of chemicals should be aided by a computer system. Its simplest form should consist of a data base accessible to the workers of a given institution (i.e. through an internet browser) and an electronic mailing system to assertain an optimal circulation of information (requests for giving over the chemicals, placing orders, and the like). The access to the system must be limited to a small circle of authorized persons employed in a given institution. Such resolution ensures easy and comfortable checking on the amounts of stored chemicals by the inquirers and placing orders to replenish the stock. This dénouement raises ergonomicity of the store functioning. Moreover, it allows for considerable reduction of a direct physical contact with chemicals of the storemans as well as the workers.

Apart from managing the store the storemans are also responsible for ordering the chemicals. Due to economic and organizational reasons it is recommended that larger batches of chemicals are ordered and delivered on specified weekdays. This increases effectiveness of work and limits staying in the store to a minimum.

Upon checking conformity of the amounts of delivered chemicals with the amounts specified in the invoices (or in other dispatch documents) the chemicals are received into the store. The structure of the data base is not strictly defined yet it should include certain essential information, i.e.
• chemical's name,
• CAS (Chemical Abstracts Service) number,
• unique id number,
• amount,
• purity,
• supplier's name and address,
• name of a group of final users.

It is recommended while creating (or buying) a data base system to anticipate a possibility of including in this system a chemical safety card (or sheet) for each chemical reagent (if such a card is required). This will enable a final user an easy access to these cards.

To make the work more efficient it is also advisable to use a system of bar codes. Integration of the bar code system with the data base reduces in a significant way a number of possible mistakes and facilitates the process of control over circulation of chemicals and their packagings within a given institution.

Figure 2. Example of the bar code system applied to a chemical compound label in the central chemical storehouse.

Depending on the character of the activities conducted by a given institution the work in the storehouse can be organized in different ways. However, it is recommended to specify if possible the days when the chemicals are received into the store, the days when the chemicals are passed onto the users and the days when the gathered waste is taken from the users and brought into the store. One day should also be devoted to the maintenance work inside the storehouse.

3.4. Passing chemicals on users

The method of giving over the chemicals should be specified by appropriate internal rules relevant to the organizational character of a given institution. As an example one can refer to the procedure described below.

Chemical reagents are passed on the users on specified days and hours. Requests for chemicals are collected through electronic mail. A request should contain a chemical's name and a shelf number on which the chemical is placed. Next, on a specified day the requests are printed and taken to the store. The chemicals are then picked up pursuant to these requests and their identification (id) numbers are written down on the printed requests. A person who placed a request checks conformity of the characteristics of the obtained chemical with the characterictics specified in the request (name, amount, purity) and after successful verification confirms with own signature a receipt of the chemical – at that moment the request becomes a pass card of the chemical. Next a storeman updates the content of the data base in accordance with the produced pass cards – at that point an invaluable assistance comes from application of the afore-mentioned system of the bar codes. One has to provide for unusual circumstances as well, for instance an urgent need of giving over the chemical dictated by additional demand, breaking of the packaging and the like. In such cases a flexible approach to the arisen problem is required and a resolution of the problem based on mutual communication of both parties, a person placing a request and a storeman. However, one has to make every endeavour to ensure that the applied special procedure has no disruptive impact on functioning of the store and first of all that it does not infringes the consistence between the information contained in the chemical data base and the actual state of chemicals in the storehouse. Also in this case in spite of the application of the special procedure the appropriate chemical pass cards have to be immediately made out.

4. Chemical Reagents at the Workplace

4.1. Formal requirements with regard to final user

Persons working with chemicals, despite possessing the factual knowledge, must be aware of numerous additional conditions that must be obeyed while using the chemicals. This means that the outsiders who did not pass appropriate training or persons not possessing appropriate factual knowledge are not allowed to be present in laboratory – laboratory can not become a place of social gatherings or a dining room.

Persons working on their own must be trained within the scope of general and occupational safety and health rules and fire regulations. The trainings should be organized on periodic terms.

In addition, there is a need of periodic medical examinations which in the chemical sector are usually required a little more often than in other fields of human activities and have broader scope (i.e. hepatic tests, allergic tests and so on).

Persons not being workers of a given institution (such as students) must be trained within the scope of occupational fire safety rules by persons being in charge of and/or conducting the laboratory activities and their work and presence in the laboratory area is allowed only and solely under the control of the supervisors. One has to spare no effort to provide these persons with all information on chemicals (i.e. chemical safety cards) that will be used by them.

4.2. Formal requirements with regard to the workplace

The question of great importance but oftentimes passed over is occupational hygiene at the workplace. At issue here are the problems of risk assessment, undertaking of activities aimed at removing and minimizing the risk, controlling the effectiveness of safety provisions and verification of the risk assessment. The most important thing in the risk assessment is taking into account hazardous properties of the used chemicals and the worker's exposure time.

In assessment of risks one has to provide for emergency situations and work out the procedures of conduct in such situations (first-aid, evacuation). The potential risks should be eliminated whenever possible. Substitutes for extremely hazardous substances must be sought and the time of the worker's contact with the chemicals must be shortened.

After each significant change at the workplace the verification of the risk assessment and applied safety procedures has to be performed anew.

Adherence to rules of occupational hygiene during activities involving use of chemicals has also its profitable economic dimension which should constitute an additional motivation to observe these rules. The benefits resulting from such behaviour are mainly bound to the worker's health (personal benefits of an employee and an employer as regards bearing the costs of medical attention and compensations), and conformity to the legally binding regulations – avoiding fines imposed by the supervising authorities. Having this in mind it is obvious that the work with extremely dangerous substances should be minimized. An example of a compound that has to be immediately withdrawn from use (replaced) is carbon tetrachloride. In APPENDICES as Appendix 2 a full chemical safety card of this chemical reagent is featured.

In each chemical laboratory room there should be a stand allocated for gathering of chemical waste. This place must meet specific requirements. It has to be arranged in accordance with the occupational safety and health rules and fire regulations and can not be located in evacuational walkways, hallways or other means of egress. In addition, it should be properly labeled – should be furnished with posters containing the information on labeling of dangerous substances and their classification.

Chemical wastes should be collected upon division into classes to containers with a seal of approval. Solid wastes are collected in polyethylene or polypropylene containers with wide openings, packed in polyethylene bags or alternatively in the producer's packaging. The containers with the wastes must be permanently and distinctly labeled in accordance with the waste classification labeling. Liquid wastes belonging to groups O, F, and S (symbols explained further on) are collected separately to canisters having capacity of 5 or 10 dm^3 made from HDPE (High Density PolyEthylene). The canisters must posses a seal of approval certifying that they can be used for storing and transportation of aggressive chemical products. Each canister must posses a legible label informing what kind of waste is placed inside the canister. The canisters should be filled up to 4/5 of their capacity and then as soon as possible passed onto the central chemical storehouse together with a protocol of the transfer of the waste substances. Description of the waste content should be also provided on the canister label. Unsuitable for chemical deactivation wastes belonging to groups TN and TP (symbols explained further on) are packed in safety, preferably unbreakable packagings, provided with permanent descriptions, placed in proper containers with wide inlets and stored in the areas designated by the instruction on storing of acutely toxic substances.

Collection and storage of wastes in unlabeled containers is strongly forbidden.

In educational processes the instruction manuals should contain a description of the method of handling the substances and wastes generated during the chemical experiments.

4.3. Chemical laboratory activities

Due to diversified character of activities conducted in specific institutions a unique description of the rules applicable during the work in a laboratory in which the chemical reagents are used is rather ambiguous. Predominantly, beside common "good laboratory practices", the internal regulations are applied that define a course of laboratory activities, emergency procedures and the like that must be observed with no excuse.

The imposed regulations take on a specific character if one deals with dangerous substances. In such cases the procedures related to storing, using and keeping records of that type of chemicals have to be clearly defined.

It is tentatively assumed that the potential users thanks to the possessed knowledge and professional experience have already got acquainted with the safety measures that must be preserved during the work with chemical substances. Since in a laboratory usually only small amounts of chemical compounds are used the serious risks for health are already escapable by a strict preservation of the recommendations left on the label provided to each chemical packaging. Every person working with chemical substances should always preserve the safety rules regardless of whether the product label bears a risk statement or not. The safety (S) rules must always be observed and in particular these that prescribe:

- to wear an eye shield incessantly and if possible to wear the gloves
- to perform as many activities as possibile under the efficiently working laboratory hood or at least wearing, in accordance with the requirements, a personal inhalation protective equipment in a well ventilated room
- to avoid any contact of the compound with the skin, eyes and mucuous membranes
- to immediately wash off the possibile sprinkles on the skin with a large amount of cold water and to apply polyethylene glycol to rinse off the lipophilic substances
- to carefully rinse out the caustic substances that penetrated into the eye with a soft stream of water (for example by focusing the stream of water onto the eye) holding the eyelids wide open and moving the eyeball in all possible ways. Immediately contact a physician providing the name of a used substance
- to promptly take off the clothing that could have got contaminated with the dangerous substances
- to seek medical advice in case of an accident or if feeling unwell
- do not eat, drink or smoke in a laboratory room
- to comply with the prescriptions contained in the chemical safety cards and standard operational procedures

The detailed compilation of the risk and safety phrases concerning the use of chemicals is presented in APPENDICES as Appendix 3.

It is always desirable to have a topical list of dangerous substances present at the workplace and to have within reach the printed chemical safety cards with the contents that could be known to every working person. In case of newly synthesized substances the risk assessment cards have to be compiled. Each such card should contain description of the safety measures that ought to be adopted while working with such a substance; the recommendations on how the substance must be packed and stored, and what should be done with the derived waste.

Substances creating the most serious hazard are liable to the detailed accounting of expenditures and to the protection against interception by the unauthorized persons. These requirements are laid down for instance on the following substances: the substances labeled as T+ (very toxic); substances labeled as C (corrosive) and additionally as R35 (causing severe burns); methanol and methanol preparations with concentrations higher than 3%; carcinogenic and mutagenic substances of category 1 or 2; intoxicating/ psychotropic preparations and their precursors labeled as I-R.

The use of the dangerous substance calls for application of the safety measures described in the relevant chemical safety card or defined by a supplier of the newly synthesized substance. Using of the substance posing a risk of exposure by inhalation is acceptable only upon evaluation that the substance concentration does nor exceed the admissible limit. Activities involving the usage of carcinogenic or mutagenic substances of category 1 or 2 are subject to rigorous monitoring in order to determine the personal exposure values for everone working with such substances and in particular: the user's exposure time (date and number of hours) and the amount of the substance used on a given working day.

As it was mentioned before only such a supply of chemical reagents is permissible at the workplace which guarantees the continuity of the chemical processes. All the excess amounts of chemicals should be stored in the central chemical storehouse. Despite that the amounts of chemicals that are indispensable at the workplace must be stored under the conditions defined in the chemical safety cards. The dangerous substances may be kept in laboratory rooms if placed in substitute (not original) packagings or laboratory containers satisfying the requirements defined in the chemical safety cards or in the professional risk assessment cards. Descriptions/labels on the substitute packagings or containers should not undergo deterioration under the laboratory conditions. This type of indelible notification must include data allowing for unique identification of a substance in question or a person who placed this substance in a provisional packaging or container.

The conducted research or educational activities may lead to generation of chemical substances in form of the vapours or gases. This is the only type of the waste that must be caught and chemically deactivated directly during its generation. To these wastes belong first of all the gases and the vapours of volatile substances that are toxic and harmful to health such as chlorine, bromine, hydrogen sulfide, hydrogen cyanide, hydrogen chloride, hydrogen bromide, phosgene, ammonia, sulphur oxides, nitrogen oxides, carbon monoxide, carbon disulfide, metal carbonyls and so forth. To this group of wastes also belong the irritating and stinking vapours of volatile organic compounds such as acrylaldehyde, alkyl acrylates, thiols (mercaptans), sulfides (thioethers) and the like as well as the vapours of diverse organic solvents such as methanol, ethanol, acetone, carbon tetrachloride, chloroform, dichloromethane, ethyl ether and aromatic hydrocarbons such as for example benzene, toluene and aliphatic hydrocarbons including pentane, hexane and the like.

The gases and the vapours of the substances that are toxic and harmful to health as well as the substances that are characterized by an unpleasant odour such as chlorine, bromine, hydrogen sulfide, hydrogen cyanide, hydrogen chloride, hydrogen bromide, phosgene, ammonia, sulphur oxides and nitrogen oxides must be absorbed in washers or scrubbers filled with proper absorbents adapted suitably to the chemical properties of the evapourating substances (for further information see the detailed subchapter on the methods of chemical destruction of different types of chemical compounds). One must not allow for release of such substances directly into the air. To absorb basic substances the solutions of sulfuric or hydrochloric acids are used preferably; to absorb acids or substances hydrolyzing in the presence of bases the solutions of sodium hydroxide or sodium carbonate are employed; to oxidize certain substances the solutions of sodium or calcium oxochlorate(I) (hypochlorite) as well as the solutions of sodium disulfate(VI) (pyrosulfate) are commonly used; to reduce some other substances the use is made of the solutions of sodium disulfate(IV) (pyrosulfite) or the alkaline solutions of sodium borohydrides. Upon completion of the reaction the absorber contents are treated such as the appropriate liquid wastes.

Also other readily vapourizing substances released from the reaction mixtures or in the course of different chemical operations (for example the solvents containing halogens or benzene) may create a serious hazard to health and a risk of fire or explosion (for example the mixtures of hydrocarbons, alcohols, ethers or acetone with air). Such substances should be liquidified using the appropriate and efficient cooling system or in particular cases absorbed on suitably chosen absorptive material.

Finally, the fissionable materials should also be mentioned. While dealing with such materials one has to adhere to the relevant regulations specifying how to behave in their presence and imposing in particular the scrupulous observance of the time spent in the laboratory and the use of the personal protective equipment. While working with the fissionable materials one has to do everything to limit to a minimum the emission of radiation to the nearest neighbourhood involving this way a harmful exposure to the radiation.

Chemical substances and preparations are divided into 15 classes in accordance with their dangerous properties (as laid down in Annex VI of the Dangerous Substance Directive 67/548/EEC). The 15 dangerous properties represent the physical-chemical properties: extremely flammable (F+), highly flammable (F), flammable (without an acronym and without a danger symbol), oxidizing (O) and explosive (E); the acute toxic properties: very toxic (T+), toxic (T), harmful (Xn), corrosive (C) and irritative (Xi); the special toxic properties: sensitizing (Xi for respiratory allergens and Xn for dermal allergens), carcinogenic (T for categories 1 and 2; Xn for category 3), mutagenic (T for categories 1 and 2; Xn for category 3) and harmful to reproduction (T for category 1 and 2; Xn for category 3); and the eco-toxic property: dangerous for the environment (N). As it can be noticed, not every dangerous property can be characterized by its own danger symbol. This stems from the fact that there are only 10 danger symbols for 15 dangerous properties. Thus the correct classification of substances or preparations cannot be deduced from the danger symbol or the indication of danger that it represents. The assignment of a substance or preparation to one of 15 classes of dangerous properties is additionally supported by the dedicated R-phrases (see the risk phrases presented in Appendix 3). The danger symbol and the wording of the indication of danger form part of the substance or preparation labeling. The detailed definitions of 15 danger classes implemented in the Polish legislation, supplemented with a danger class for substances creating a biological hazard and a danger class for radioactive substances, are presented in Table 1. This table actually contains a compilation of the introduced danger characteristics with chemical examples, the dedicated acronyms and danger symbols (pictograms).

Table 1. Classification of dangerous substances and preparations with descriptions and examples in accordance with the Regulation of the Ministry of Health of 2 September 2003 (*JL 2003 No. 171, item 1666, JL 2004 No. 243, item 2440 and JL 2007 No. 174, item 1222*).

Class	Characteristics	Examples	Acronym	Symbol
Explosive substances	**Criteria:** There is a possibility of explosion due to shocks, collisions, friction and contact with air. These compounds and preparations can react egzothermally in the environment void of atmospheric oxygen. **Safety measures:** Avoid collisions, shocks, friction, sparks, fire and heat.	ethyl trioxonitrate(V), lead(II) azide, (di)benzoyl peroxide, picric acid, 2,4,6-trinitrotoluene (TNT), mercury fulminate, glyceryl trinitrate	E	
Oxidizing substances	**Criteria:** Either flammable or able to catch fire on contact with combustible materials. Through oxygen release significantly increase a risk of fire, favour its spreading and hinder its extinguishing. **Safety measures:** Avoid any contact with combustible substances.	hydrogen peroxide, potassium tetraoxomanganate(VII), trioxonitric(V) acid, chromium(VI) oxide, potassium tetraoxochlorate(VII)	O	
Flammable substances	**Criteria:** Liquids with a flash point from 21°C to 55°C. **Safety measures:** Keep away from unshielded fire, sparks and sources of heat.	butanol, pentanol	No acronym	No symbol
Highly flammable substances	**Criteria:** Substances flammable in air upon heating, solids which upon short contact with a source of ignition may catch fire, liquids with a flash point below 21°C and substances which on contact with water emit flammable products. **Safety measures:** Keep away from unshielded fire, sparks and sources of heat.	benzene, ethanol, white phosphorus, sodium, calcium carbide, metal hydrides, methyl lithium, silanes, very fine metal dusts e.g. aluminium dust	F	
Extremely flammable substances	**Criteria:** Gases flammable under normal temperature and pressure and liquids with a flash point below 21°C and a boiling point not exceeding 35°C. **Safety measures:** Keep away from unshielded fire, sparks and sources of heat.	alkanes from C_1 to C_4, ethyl chloride, hydrogen, carbon monoxide, diethylether, vinyl chloride, acetaldehyde, ethylamine	F+	
Toxic substances	**Criteria:** Toxic to human's organism if swallowed, inhaled or absorbed through skin. **Safety measures:** Avoid any bodily contact. If feel unwell immediately contact a physician.	sodium fluoride, potassium fluoride, ammonia, chlorine, methanol, phenol, acetonitrile, bromofor, carbon tetrachloride, aniline	T	

Class	Characteristics	Examples	Acronym	Symbol
Very toxic sub-stances	**Criteria:** Very toxic if swallowed, inhaled or absorbed through skin even in trace doses, may cause serious irreversible damage to health. **Safety measures:** Avoid any bodily contact. If feel unwell immediately contact a physician.	nitrogen dioxide, hydro-cyanic acid and its salts (excluding complexes), mercury compounds, hydrogen fluoride, hydrogen sulfide, bromine, phosgene, acrolein, nitrobenzene	T+	
Corrosive substances	**Criteria:** Substances causing corrosive effect, prolonged or repeated exposure of the skin or mucuous membranes to these substances may result in inflammations. Risk of sensitization by skin contact (if classified as R43). **Safety measures:** Appropriate safety measures must be undertaken to protect eyes, skin and clothing. Do not breathe va-pours! In case of accident or if feel unwell immediately contact a physician.	sodium hydroxide, potassium hydroxide, trioxonitric(V) acid, hydrogen fluoride, sulfuric acid, phosphoric acid, antimony pentachloride, formic acid, acetic acid, propionic acid, triethylamine, benzylamine	C	
Sensitizing substances	**Criteria:** Sensitizing effects through skin or respiratory tract contacts. Respira-tory allergens are labeled with the danger identification Xn (harmful) while dermal allergens are given the danger identifica-tion Xi (irritant). **Safety measures:** Avoid skin contact, do not inhale vapours.	**respiratory allergens:** cobalt, nickel (II) sulfate, maleic acid anhydride, phthalic acid anhydride, isocyanate, glycidole **skin contact allergens:** hydrazine, formalde-hyde, alkyl acrylates, cyanamide	Xn	
Irritant substances	**Criteria:** Cause skin inflammation, irrita-tive effects or impair the respiratory system and eyesight. **Safety measures:** Avoid eye and skin contacts, do not inhale vapours	calcium chloride, sodium carbonate, dimethylamine	Xi	
Harmful substances	**Criteria:** Substances which can have a harmful effect on human's health or cause a risk of irreversible damage to health. Risk of sensitization by inhalation (defined as R42). **Safety measures:** Avoid any bodily contact.	iodine, manganese dioxide, potassium chlorate, methylene chloride, chloroform, oxalic acid, hydroquinone, glycol, toluene, benzaldehyde, aminophenol	Xn	

Class	Characteristics	Examples	Acronym	Symbol
Carci-nogenic substances	**Criteria:** Substances for which there is a clear evidence of activation of tumors (category **1**) or these which with a high probability should be regarded as car-cinogens (category **2**) or which are just suspected of having a possible carcinogenic potential (category **3**). Carcinogenic substances belonging to category **3** are labeled with the acronym Xn and the danger symbol characteristic of the harmful substances (St. Andrew's cross). **Safety measures:** Avoid any bodily contact.	category **1:** benzene, benzidine and its salts, 2-naphthylamine and its salts, vinylchloride, aminobiphenyl, butadiene, asbestos, arsenic trioxide, nickel dioxide, zinc chromate category **2:** 3,4-benzo[a]pyrene, o-toluidine, hydrazine, acrylonitrile, acrylamide, diazomethane, 1,2-dichloroethane, beryllium and its compounds category **3:** acetamide, aniline and its salts, chloroform, carbon tetrachloride, 1,4-dioxane, DDT, lead chromate, nickel hydroxide	T Xn	
Mutagenie substances	**Criteria:** Substances causing inheritable genetic damages. To category **1** belong the substances for which there is a sufficient evidence of mutagenic effect, to category **2** these which with a high probability should be regarded as mutagens and to category **3** those which are suspected of mutagenic potential. Substances classified as muta-genic category **3** have to be assigned the danger symbol St. Andrew's cross and the danger identification harmful (Xn). **Safety measures:** Avoid any bodily contact.	category **1:** up to now, no substance has been detected which has such an effect on man category **2:** acrylamide, benzo[a]pyrene, diethylsulfate, ethyleneoxide, 1,2-dibromo-3-chloro-propane category **3:** methyl thiophenate, zinc bis(dimethyldithio)-carbamate, benomyl, atrazine, phosphamidon	T Xn	

Class	Characteristics	Examples	Acronym	Symbol
Substances harmful (toxic) to reproduction	**Criteria:** Substances which can impair (a) fertility or which can damage (b) the development of the unborn child. Substances which are suspected of impairing the fertility or of having developmental damaging properties belong to category **3** and are labeled with the acronym Xn and the danger symbol St. Andrew's cross. **Safety measures:** Avoid any bodily contact.	category **1** (a): up to now, no chemicals impairing fertility are classified in this category category **1** (b): majority of lead compounds, methylmercury, carbon monoxide category **2** (a): benzo[a]pyrene, 4-chlorobenzo-trichloride, polychlorobiphenyls (PCBs), lead hexafluorosilicate category **2** (b): dinitrophenols, glycolethers, dimethylformamide category **3** (a): lead chromate, nitrobenzene, nitrotoluene category **3** (b): carbon disulfide, α-chlorotoluene	T Xn	
Substances dangerous for the environment	**Criteria:** Release to aquatic or non-aquatic environment would or could cause an immediate or delayed damage to one or more environmental constituents which might lead to immediate or delayed environmental equilibrium disturbance. Some substances or products of their decomposition may simultaneously affect different environmental elements. **Safety measures:** Depending on a scale of risk do not allow for penetration into the water system, soil and environment. Adhere to relevant rules concerning the removal.	**R50** substances: DDT, ammonia, pentachlorophenol, aniline,chloroacetic acid, halogens, hydroquinone, p-benzoquinone, copper sulfate, disulfur dichloride **R51/53** substances: N,N-diethylaniline, nitrotoluene, thiourea, benzotrifluoride, 2,4-dichlorophenol **R52** substances: methyl chloride, hexane, chloroacetonitrile, methylcyclohexane, chlorobenzene **R58** substances: lead(II) methanesulfonate, 3-chlor-4,5-α-pentafluorotoluene **R59** substances: carbon tetrachloride, 1,1,1-trichloroethane, bromomethane	N	

Class	Characteristics	Examples	Acronym	Symbol
	Classes which are not specified in the Regulation but which are oftentimes included in the lists of substances and preparations with the assigned danger symbols			
Substances creating biological hazards	**Criteria:** Living organisms or substances of organic origin that pose a hazard to human health. This may include the medical waste, i.e. micro-organisms, viruses or toxins (of biological origin) that may constitute a threat to human life. Biological hazard may also arise from substances posing hazard to animals. **Four Levels of Hazard:** **1.** – Beneficial bacteria and viruses of *Bacillus substilis, Escherichia coli,* and *Varicella* genus, that do not cause infection. At this level the biological hazard is at minimum. Such wastes are usually removed with other residues (though their separation is recommended). Decontamination procedures at this level are consistent with similar modern methods of virus removal. In laboratory environment all the materials having contact with bacteria and/or bacteria cultures are cleansed in an autoclave. **2.** – Different bacteria and viruses with a mean infectiousness for humans or such that are hardly neutralized in laboratory using an aerosol, i.e. WZW viruses of the type A, B, or C, virus of influenza of the type A, virus of Lyme disease, salmonella, virus of mumps, virus of measles and HIV. **3.** – Bacteria and viruses that may cause heavy, fatal diseases in humans but against which there are no effective vaccines, i.e. anthrax, West Nile fever, SARS, pox, Rift valley fever, blotchy fever and jaundice. **4.** – Viruses that may cause heavy diseases in humans against which there are no available vaccines, i.e. Argentinian bleeding fever, viruses of Ebola, Lassa and Marburg.		No acronym	
Radioactive substances	**Criteria:** Radioactive substances resulting from processing (production) of nuclear fuel, separation and application of radioactive isotopes. **Safety measures:** Observe in absolute terms the relevant regulations. Seeing this or a similar danger symbol better do not approach the danger zone if unnecessary.	uranium, polonium and other isotopes with a short half-life	No acronym	

5. Gathering and Preparation of Chemical Wastes and Packagings for Recovery and Disposal

5.1. Classification of the waste at the workplace

Chemical wastes as well as the glass and plastic packagings gathered at the workplace have to be periodically passed on the central chemical storehouse and then transferred from there to a specialized off-site facility for recovery and disposal.

Some fraction of the generated waste belongs to the dangerous wastes therefore all the generated wastes are treated with special care as dangerous wastes or potentially dangerous wastes and are handled in accordance with the instructions drawn up for the dangerous wastes.

As a dangerous substance is regarded each single substance or each mixture of substances that due to its chemical, biological or radioactive properties if unproperly handled may cause a hazard to health or life of humans or may lead to environmental pollution; the dangerous substance may be a raw material, product, intermediate product, waste and a substance that came into existence as a result of breakdown of a particular device or equipment.

Municipal wastes, i.e. waste paper, packagings made from glass and plastics, metal scrap and discarded equipment except the waste mercury being a part of it are not rated among chemical wastes.

According to Art. 10 and Art. 17 of the Act on packagings and packaging waste of 11 May 2001: "A user of the dangerous chemical substances is obliged to return to a seller the remaining multiple-use packagings and packaging waste". It should be stressed that in accordance with the above cited articles no all packagings are required to be returned to a seller. This concerns only the packagings and packaging waste containing previously the following used up substances/preparations
– very toxic,
– toxic,
– carcinogenic of category 1 or 2,
– mutagenic of category 1 or 2,
– harmful to reproduction of category 1 or 2,
– dangerous for the environment.

A problem may arise, however, for an unaware user how to recognize whether a given packaging should be returned to a seller or not.

In regulation of the Ministry of Health of 2 September 2003 (*JL 2003 No. 171, item 1666*) the detailed criteria and method of classification of chemical substances and preparations have been defined according to which the substances/preparations classified as very toxic, toxic, carcinogenic of category 1 and 2, harmful to reproduction of category 1 and 2, and harmful for the environment have the following labeling:
– Very toxic substances/preparations: the acronym T+ and at least one of the labels R26, R27, R28, R39, R39/26, R39/27, R39/28, R39/26/27, R39/26/28, R39/27/28 or R39/26/27/28.
– Toxic substances/preparations: the acronym T and at least one of the labels R23, R24, R25, R39, R48, R39/23, R39/24, R39/25, R39/23/24, R39/23/25, R39/24/25, R39/23/24/25, R48/23, R48/24, R48/25, R48/23/24, R48/23/25, R48/24/25 or R48/23/24/25.
– Carcinogenic substances/preparations of category 1 or 2: the acronym T and at least one of the labels R45 or R49.
– Mutagenic substances/preparations of category 1 or 2: the acronym T and the label R46.
– Substances/preparations harmful to reproduction of category 1 or 2: the acronym T and at least one of the labels R60 or R61.
– Substances/preparations dangerous for the environment: the acronym N and at least one of the labels R50, R52, R53, R54, R55, R56, R57, R58, R59, R50-53, R51-53 or R52-53.

The rules of dealing with the chemical waste generated in chemical laboratories should be consistent with the basic legal acts: the act of 27 April 2001, "The environmental protection law" (*JL 2001 No. 62, item 627*), the act of 27 April 2001, "Act on waste" (*JL 2001 No. 62, item 628*) and the legal regulations adopted by the voivodeship and local authorities.

Gathering of the waste apart from a specific method of the waste segregation, sorting and labeling requires that several general rules have to be obeyed. Reception and gathering of the chemical waste must be conducted in conformity with the occupational safety and health rules and the fire regulations. To the central chemical storehouse only the prepared (packed) waste is taken so it would not cause a hazard during receiving, storing and the future waste management.

The waste destined for recovery and disposal, taking into account its composition and physical state, is divided into the following classes:

Table 2. Classification of chemical wastes.

Labeling	Waste composition and properties
O	liquid, organic, without halogens
F	liquid, organic, containing halogens
P	combustible, solid
N	incombustible, solid
S	saline solutions, pH=6–8
TN	very toxic, incombustible
TP	very toxic, combustible
R	mercury and mercury compounds
Various	regenerable organic solvents (containing at least 80% main solvent in the waste)

5.2. Dealing with the chemical waste

Classification of the waste determines a method of the waste storage. Below the methods of dealing with particular waste classes are characterized. These characterizations may be considered as the workstation instructions for persons directly responsible for receiving and gathering of the chemical waste.

5.2.1. Dealing with the waste classified in groups O and F

Liquid wastes classified as belonging to groups **O (organic without halogens)** and **F (organic containing halogens)** should be collected separately in packagings with capacity of 10 l made from HDPE (High Density PolyEthylene). The containers must have a seal of approval certifying that they are suitable for storing and transporting of aggressive chemical products.

In addition, the wastes of the type **O** cannot contain more than 3% halogens (altogether). Among the wastes belonging to the group **F** there may be organic compounds not containing halogens. This division results from the fact that the chemical substances containing halogens (unless not recovered from the waste) should be incinerated separately since in the course of their incineration under inappropriate conditions extremely toxic dioxins are formed. The total content of halogens higher than 3% decides on the assignement of a mixture to the group **F** in the waste classification. The wastes containing more than 3% halogens may be rated as belonging to the group **O**.

Sort of a variation of the above described chemical wastes are mixed solvents containing at least 80% one single solvent **(Various)**. It is recommended to collect them separately. Such mixtures are quite suitable for recycling and recovery of the main component.

5.2.2. Dealing with the waste classified in groups P and N

Each solid waste assigned to groups **P** (solid combustible) or **N** (solid incombustible) is packed in plastic bags made from polyethylene or in original unbreakable producer's packagings. Then each bag or packaging is provided with an indelible description of its content and placed in a container made from HDPE with a wide opening. Each type of the waste is gathered in a separate set of containers.

5.2.3. Dealing with the waste classified in groups TN and TP

The wastes labeled with the acronym **T** containing toxic organic as well as inorganic substances **have to be deactivated chemically before placing them in the containers for overall collection of the waste.** Deactivation should be performed with a greatest care in accordance with the rules described in the instructions included in the chemical safety cards. Negligence of the deactivation duty may cause a danger to health or life through accidents that may happen in a laboratory room where these wastes have been generated and are temporalily stored as well as at further stages of the waste management aimed at the waste recovery and elimination.

In unique cases when both organic and inorganic toxic substances cannot be deactivated they are collected in separate containers. The waste containing such substances must be placed in a tight and unbreakable packaging supplied with a clear description of its content and stored in containers labeled with the symbols **TP** or **TN**. The rooms for storing the waste containers labeled as **TP** and **TN** should be selected in accordance with the instructions on how to store the toxic substances. Access of the outsiders to these containers should be made impossible.

5.2.4. Dealing with the waste classified as the type S (saline solutions with pH=6–8)

Liquid residues classified as the waste of the type **S** must be systematically checked with respect to their acidity. These residues should be neutralized so as to place pH within the range 6–8. From the solutions containing cations of heavy metals the hydroxides or sulfides of these metals should be precipitated. Insoluble precipitate must be separated from the suspension by decantation and filtering off, packed in polyethylene bags, placed in tight and unbreakable packagings provided with durable descriptions of the contents and stored in the containers labeled with the symbol **S**.

5.2.5. Dealing with the waste containing mercury, group R

Chemical wastes containing mercury assigned to the group **R** belong to the wastes that are exceptionally toxic and dangerous for the natural environment. Therefore they are distinguished as a separate group of wastes. These wastes are gathered in separate containers and labeled with the symbol **R**.

The spent metallic mercury coming from the unused equipment, broken thermometers, manometers or other devices must be carefully gathered, separated from mechanical contaminations and placed in tight and safety packaging. It must not be stored any longer in the laboratory rooms for there is a risk of inhalation of the mercury vapours which are toxic.

The spilled mercury should be collected very carefully and dealt with in the same fashion like the mercury from the discarded equipment. Other mercury residues should be solidified using a zinc amalgam, treated with sulphur or absorbed by using special brand-name preparations such as for instance Chemisorb Hg, packed in polyethylene bags and stored in the containers marked with the label **R**.

Solutions containing mercury salts such as mercury acetate or mercury chloride must be devoid of mercury cations by binding them to the ion-exchange resin. Very diluted aqueous solutions containing the Hg^{2+} cations can be devoid of these cations by reduction to metallic mercury in an iron dust packed column.

5.2.6. Dealing with the radioactive wastes

A facility engaged in Poland in processing of radioactive wastes is the Radioactive Waste Management Plant (RWMP) having its seat in Świerk near Otwock.

Liquid low level wastes are purified by sorption; in this process over 99% radio-nuclides gets removed from the sludges deposited on the sorptive material. Then this material as a radioactive cancentrate is incorporated into asphalt.

Liquid intermediate level wastes are subject to evaporation. The evaporation residues contain over 99.9% radioactive substances which are then solidified in concrete.

Filter cardridges from the systems of water purification in the cooling water circuits of the nuclear power reactors are stabilised by epoxide resins.

Solid low level radioactive wastes are compacted in a hydraulic press. Due to this, depending on a character and properties of the waste, its volume is reduced 3–5 times. Solid and solidified radioactive wastes are placed in steel drums protected from corrosion by a layer of zinc and then transported to a waste disposal facility. The wastes are transported using dedicated, purpose-built transport vehicles, taking all appropriate precautions.

Worldwide in many laboratories the work on the methods of rendering the radioactive wastes innocuous is now in progress. Especially promising seems to be so called transmutation or transformation of long-lived radioactive wastes into substances (radionuclides) with significantly shorter half-lifes and thus radiologically less active and hazardous.

In Poland, likewise in other countries that do not have nuclear power plants, the radioactive wastes are disposed of in shallow earth burial series.

6. Waste Transfer for Recovery and Disposal

The chemical wastes collected and packed in the central chemical storehouse have to be transferred on periodic terms to a waste treatment plant. Usually it takes place through the transfer of the waste to an off-site facility specialized in the waste treatment and disposal.

6.1. Choice of the waste transfer company

The crucial point at this stage is the choice of the proper waste treatment company. First of all, the selected waste management company must posses all the required permits as regards its functioning. Second, it should ensure a possibility of receiving as broad spectrum of wastes as possible at a price economically reasonable from the point of view of the waste holder. There are several waste treatment companies on the market but oftentimes these companies posses permits for processing only the limited range of wastes or the charge for recovery and disposal of the extremely dangerous waste is exorbitantly high (usually in this case these companies act just as a go-between). As regards these conditions one must make every endeavour to be sure that the choice of the waste management company is actually a result of a full market analysis. It should be mentioned that the distance factor in Polish conditions is rather negligible since the costs of transportation are incommensurately low as compared to the costs of the treatment of the chemical wastes and packagings. Upon making selection of a few companies the terms of cooperation must be very precisely fixed. The waste reception frequency, amounts of the waste in particular batches and the ways of the waste preparation to the transfer must be strictly defined. Upon making the final choice of the waste management company it is recommended that a relevant formal contract should be prepared and signed by both sides.

6.2. Classification of wastes transferred for recovery and disposal

Prior to recovery and disposal the information on the quantitative and qualtitative composition of the waste classified with the waste codes pursuant to the regulation of the Ministry of Environment of 27 September 2001 (*JL 2001 No. 112, item 1206*) is provided to the waste management facility. On the basis of these data the waste transfer cards are compiled which are indispensable during transportation and become the waste transfer documents – pursuant to the regulation of the Ministry of Environment of 14 February 2006 on specimens of documents used for needs of waste accounting (*JL 2006 No. 30, item 213*).

Taking into account various sources of the waste generation the wastes are divided, in accordance with the above mentioned regulation, into 20 groups:
- wastes resulting from exploration, mining, quarrying, physical and chemical treatment of ores and other minerals – 01,
- wastes from agriculture, horticulture, aquaculture, forestry, hunting and fishing, food preparation and processing – 02,
- wastes from wood processing and the production of panels and furniture, pulp, paper and cardboard – 03,
- wastes from the leather, fur and textile industries – 04,
- wastes from petroleum refining, natural gas purification and pyrolytic treatment of coal – 05,
- wastes from the manufacture, formulation, supply and use of the products of inorganic chemical industry – 06,
- wastes from the manufacture, formulation, supply and use of the products of organic chemical industry – 07,
- wastes from the manufacture, formulation, supply and use of coatings (paints, varnishes and vitreous enamels), fillers, adhesives, sealants and printing inks – 08,
- wastes from the photographic industry and photographic services – 09,
- wastes from power stations and other combustion plants (except 19) – 10,
- wastes from chemical surface treatment and coating of metals and other materials (e.g. galvanic processes, zinc coating process, pickling processes, phosphating, alkaline degreasing, anodising) – 11,
- wastes from shaping and physical and mechanical surface treatment of metals and plastics – 12,
- oil wastes and wastes of liquid fuels (except edible oils and those in chapters 05, 12 and 19) – 13,
- waste organic solvents, refrigerants and propellants (except 07 and 08) – 14,
- waste packaging; sorbents, wiping cloths, filter materials and protective clothing not otherwise specified – 15,
- wastes not otherwise specified – 16,
- wastes from construction, renovation and demolision of buildings and from road infrastructure (including excavated soil and stones from contaminated sites) – 17,
- wastes from human and animal health care – 18,
- wastes from waste management facilities, off-site waste water treatment plants and the preparation of water intended for human consumption and water for industrial use – 19,
- municipal wastes including separately collected fractions – 20.

A full list of the waste codes is presented in APPENDICES as Appendix 4.

6.3. Collection of waste to the central chemical storehouse

Collection of chemical waste due to its specific character has to be performed in accordance with the lucid internal procedures. Owing to these, the rules of proceeding in emergency situations are clear, and the collecting and storing of the waste becomes itself more systhematic and predictable.

Some general rules should be observed:

- Each organizational unit is reponsible for the waste produced by itself, and in particular for proper waste collection and storage, for drawing up a waste transfer protocol according to the actual state and passing the waste on the central chemical storehouse.
- In each organizational unit a plenipotentiary and the storemans (as many as required) responsible for proper waste handling are appointed by a chief Manager or Director.
- The on-site generated waste is gathered in the places especially destined for this purpose. These places must meet the requirements set by the occupational safety and health rules and the fire regulations.
- The substances which are very toxic, carcinogenic, explosive, pyrophoric, acutely irritating, stinking, constituting a risk of strongly exothermic or even explosive reactions with other substances can not be placed directly in the containers for overall collection of the waste. These substances must be first chemically processed (if possible in cooperation with a local service responsible for the waste recovery and disposal) into the substances not constituting a serious hazard to the laboratory staff, the service for the waste recovery and disposal, and the natural environment. The obligation implying the waste deactivation is imposed on a unit generating the waste.
- Handling the very toxic and dangerous waste calls for a special care and an absolute fulfillment of all the occupational safety and health rules concerning the handling of acutely toxic substances. Such substances or mixtures containing such substances must not be directly placed in the overall containers for collecting the waste. Before placing them there these chemicals must undergo a complete chemical deactivation.
- Collecting and storing of the flammable organic waste requires resorting to the fire security regulations.
- All the subunits are required to pursue the activities minimizing the amount of the generated waste.
- All the waste is received in the central chemical storehouse only with an accompying waste transfer protocol including: the waste composition and waste amount, a name of a person responsible for proper description of the waste ingredients, a symbol of a subunit where the waste was produced, a date of the waste generation.
- Potential costs of the waste recovery and disposal are secured in advance in the unit budget.
- At least once a month the collected waste should be passed on the central chemical storehouse.
- Plenipotentiaries acting on behalf the subunit managers are under obligation to keep a registry of the generated waste and the documentation on passing the waste on the central chemical storehouse, whereas the Manager of the central storehouse is obliged to keep records of the received waste and the waste transferred for recovery and disposal.

Responsibility and the range of activities of each functionary should be clearly defined. Such approach increases the vigilance and additionally motivates the personnel to stick to the procedures.

The manager of the central storehouse is bound to organize and control the activities conducted in the central storehouse. In the case of collecting, recovery and disposal of the waste, the manager is also responsible for keeping the documentation concerning the taking over, recovery and disposal of the waste, the control of the related costs, preparation of the annual report fot the unit manager, surveillance over the equipment and state of the on-site waste stores (the places in subunits where the waste is collected).

The manager of the central storehouse (or an authorised person) receives the waste from the plenipotentiaries of the subunits. The containers with the waste substances are received only if the accompanying label is filled in and the waste transfer protocol is provided.

The manager of the central chemical storehouse (or an authorised person) is obliged to keep the "Waste Records" using for instance a typical pattern of the waste transfer card shown below:

Table 3. Format of the waste record card kept in the central chemical storehouse.

No./unit	Person passing the waste on	Waste symbol	Waste composition	Waste weight	Signature of the person passing the waste on	Date of the waste receipt

In addition, the manager of the central storehouse keeps the documentation concerning the recovery and disposal of the waste, and in particular this person prepares and keeps the "Waste Transfer Cards" – a typical pattern of such a card is included in Appendices as Appendix 5.

The plenipotentiaries from the internal subunits are the persons possessing a higher chemical education, designated by a manager of each internal subunit and responsible for proper and safe collection of the waste in a given subunit. In particular these persons: control the on-site waste stores and care for the equipment collected in these places, receive the waste delivered by the subunit workers, decide on a type of the collected waste (prepare the containers for collecting the waste and care for the proper container labeling), control the observance of the occupational and safety rules at work with the wastes, supervise the gathering and passing the waste on the central waste storehouse, keep the protocols of the waste transfer. Every single container with the collected waste must be properly and precisely labeled. While passing the waste on the central storehouse the filled in protocol of the transfer of the waste substances must be attached. An example of such a protocol is shown below.

Table 4. Format of a typical protocol of the waste transfer from an organizational subunit to the central chemical storehouse.

No.	Person passing the waste on	Waste symbol	Waste composition	Waste weight	Signature of the person passing the waste on	Date of the waste generation	Date of the waste receipt

The protocol drawn up on a special questionnaire must include: composition of the waste placed in the container, name and a signature of the person responsible in a given group for the waste management, name of the subunit passing the waste on, the date of the waste generation, the weight of the waste placed in the container.

The same data should be entered on a label sticked to the container filled with the waste, and if not all the data then at least those informing on a number of a subsequent container, the symbol of the waste and its weight.

7. Summary

A full migration pathway of a chemical reagent in a scientific or educational institution, starting from purchase of the reagent through its presence at the workplace till a moment of disposal of its derivative products, has been traced. On that route the legally binding regulations concerning the purchase and usage of the chemicals as well as the recovery and disposal of the chemical waste were taken into account.

The present title is meant for the persons employed in all-type institutions conducting research or educational activities where a use is made of chemical reagents.

One workable solution concerning the management of chemicals and the derived waste has been proposed. Of course, this solution may be subject to some modifications, important from a practical point of view, depending on specific conditions existing in a given institution. Introducing such modifications, one has to make every endeavour to ensure that the changes do not stand in contradiction to the established law – hence a broad chapter devoted to the acts and regulations with references to the Journal of Laws of the Polish Republic or the Official Journal of the European Community aimed at making easier the process of moving in an impenetrable maze of regulations, acts and directives.

The currently used classifications and labeling of the chemical reagents and wastes in Poland and the European Community have been presented. It is to be noticed, however, that there are far more solutions in this field. As an example one can take the classifications and labeling used in the United States or more locally that used by some European producers of chemical reagents such as for instance the Fluka company. In the case of the Fluka company the following two examples can be given: F4 – do not heat above xx°C (xx – from 1 to 15, this value must be multiplied by 10 to give the temperature in the Celsius scale), and F32 – avoid contact with the salts of heavy metals. Fluka makes use of 36 own labelings – all can be found on the Fluka web-pages.

The management of chemicals and derived chemical waste is a very challenging task. Therefore in this study a particular emphasis was placed on the selection criteria that can be applied to choose the right persons responsible for safety and health of all the people involved in the migration of the chemicals within a given institution at each stage of this migration. The efforts were undertaken to aware a user of a wide range of real dangers accompanying the processes connected with handling and storing of the chemical substances. On the basis of the presented classification the attempts were made to describe the specific behaviour and work patterns referring to the particular chemical classes taking into account from one side the potential risks and minimizing from the other side the imposed restrictions and requirements that must be met in order to make the work safe and harmless to everybody's health. At this point it should be reminded that the basic source of information on a given chemical reagent is its chemical safety card (or sheet) which should be always easily available. For a prompt preliminary risk assessment it is recommended to get aquainted as well as possible with the danger symbols collected in Table 1; an ideal solution would be hanging of a relevant poster on the wall in each laboratory room.

An important question that should be mentioned is a discrepancy in nomenclature observed in the classifications provided by Polish and the European legislations. It is not a rare occurrence that the Polish classifications are more extended and thus more accurate. This concerns for instance the chemical waste. In European classifications a whole class of the waste is sometimes defined succinctly by a single word or few words while the corresponding Polish definition is split into several subclasses with a very precise description of each. Moreover, in the Polish legal regulations there is a full range of codes which are not present whatsoever in the European legislation.

Finally, it is worth to mention that the quoted legal regulations convey a state of law in May 2008. The present study should not be considered as a source of legal acts but only as a guide and cross reference to the existing legislation.

Acknowledgment

The authors would like to thank the Faculty of Chemistry at Jagiellonian University for providing a financial support of this project.

8. Literature

Stanley E. Manahan, *Toxicological Chemistry and Biochemistry,* 3rd ed., CRC Press LLC, New York, 2003.

Czesława Rosik-Dulewska, *Principles of Waste Management* (in Polish), PWN, Warszawa, 2006.

Krzysztof Czekierda, *Dictionary of Environmental Protection. Waste Management* (in Polish), Kanion, 2007.

Wastes and Packagings – New Regulations and Requirements (in Polish), joint publication, Ed. Włodzimierz Urbaniak, Forum, Poznań, 2004.

Tony Hare, *Toxic Waste, Polluting the Air, Habitat Destruction*, Almapress, 1999.

Herbert F. Bender, Philipp Eisenbarth, *Hazardous Chemicals*, Wiley-VCH, Weinheim, 2007.

http://isap.sejm.gov.pl

http://eur-lex.europa.eu

9. Appendices

9.1. Appendix 1 – Factual and Formal Requirements of the Chemical Safety Card

1. Identification of the substance/preparation

1.1 Identification of the substance or preparation.

1.2 Potential usage of the materials/products. Character or the recommended usage of the substance/preparation should be given. This point also may contain ashort description indicating particular usage.

1.3 The name of the legal person/institution/company responsible for placing the substance or preparation on the market within the European Community has to be given. Full address and telephone number of the manufacturer, importer or distributor should be provided.

1.4 An emergency telephone number of the company and/or appropriate official advisory body responsible for providing information on possible harmfulness of the substance/preparation to human health must be given.

2. Composition/information on ingredients

The information given should allow the person making use of it to readily determine the degree of the hazard resulting from the contact and usage of the separate components of the preparation. The names and numbers of the indicated substances should be given in accordance with the EINECS (European INventory of Existing Commercial Substances) catalogue or alternatively as the IUPAC names (if available) and CAS (Chemical Abstracts Service) numbers. The substance classification with its hazard symbols and R-phrases indicating the nature of the risks should also be provided.

3. Hazard identification

The potential risks for humans and the environment that may result from the usage of the substances and preparations should be clearly and concisely indicated. A description is also necessary of the possible occurrence of the most important physicochemical consequences and manifestations harmful to humans and the environment that can be at least partly predicted as regards the usage and improper application of the substances and preparations. A need may arise to take care of other dangers that are not included in the hazard classification but may contribute to the occurrence or the increase of the general danger attributed to the usage of a given substance.

4. Measures of first aid

The information on the first aid measures should be short and understandable so that the casualty, the persons working in his/her neighborhood and the first aid helpers could easily comprehend it. This information should be collated under different titles according to the risk types such as inhalation, skin or eye contact, ingestion.

5. Fire-fighting measures

The fire-fighting measures appropriate in the case of a fire initiated by the substances and preparations in question or a fire initiated in their vicinity should be provided. There is a need to specify the suitable extinguishing media, extinguishing media which must not be used for safety reasons, special exposure hazards arising from the substance and preparation itself, combustion products, resulting gases, special protective equipment for fire-fighters.

6. Measures in case of accidental release

Depending on a given hazardous substance or preparation it may be necessary to give a specific information on personal precautions, environmental precautions and the methods of cleaning up the environment.

7. Handling and storing

The information should contain recommendations for protection of health, safety and the environment. An assistance should be given to an employer in designing the appropriate work conditions and organizational regulations.

7.1. Handling
Safety precautions for secure handling of the substance including the recommendations concerning the usage of the suitable technical equipment should be given.

7.2. Storing
The conditions of safety storage of the substance should be specified. If possible the recommendations concerning the maximum storage quantity under the existing storage conditions should be mentioned. Especially important are instructions related to individual special circumstances such as for instance the choice of the adequate package materials suitable for packing and storing of a given substance or preparation.

7.3. Specific use

In the case of end products designed for a specific use the recommendations should refer to the intended purpose of the use. These recommendations should be useful and detailed. If possible a reference to the specific already approved instructions in industry or in other domain should be given.

8. Control of the exposure of humans and the environment to chemical substances and the personal protective equipment

8.1 Exposure limit values

The presently used accurate control parameters of the occupational exposure and their limit values as well as the existing biological limit values are to be listed. These values should be given to a legal person introducing the substance or preparation on the market in a given EU member state. The information on currently recommended monitoring procedures should be provided.

8.2 Human exposure control

The exposure control includes a full range of adequate protection and prevention measures applied to reduce the exposure of the employees to the effects of the used substance or preparation. It requires designing of the appropriate work processes and engineering controls, the use of adequate equipment and materials, the application of collective protection measures at the source of exposure and eventually the personal protection measures such as for instance the personal protective equipment. If the personal protection is necessary it should be precisely specified which protection measure provides the best protection to the respiratory system, hands, eyes and skin.

8.3 Environmetal exposure control

The information essential to the employer to fulfill the existing Community environmental protection legislation has to be included.

9. Physical and chemical properties

All the available data and relevant information on the substance or preparation have to be provided (i.e. description of its physical state, colour, odour, boiling point, flash point, flammability, relative density, solubility in water/oil, viscosity, pH, and so on).

10. Stability and reactivity

It is necessary to know and avoid the conditions and materials which may cause a dangerous reaction and produce hazardous decomposition products.

11. Toxicological information

Concise but complete and comprehensible description of toxicological effects that can occur on normal handling and use must be provided. This includes the effects of the exposure to hazardous chemicals known partly from practical experience and partly form the research, the information on the different possible exposure routes (respiratory system, digestive system, skin and eyes contact), as well as the symptoms which could be related to physical, chemical and toxicological properties.

12. Ecotoxical information

The possible effects, behaviour and environmental fate of the substance or preparation in air, water and/or soil have to be mentioned. If possible the important research results on their ecotoxicity, persistence and biodegradability, potential for bio-accumulation should be given and all the relevant and available data on any adverse effects in the environment should be collected.

13. Disposal considerations

For each substance or preparation as well as each contaminated packaging the appropriate methods of disposal (incineration, recycling, etc.) must be described.

14. Transport information

All the possible special precautions which a user or a carrier must know in the case of transportation of the substance or preparation in and out of the regional area and the information such as: the UN number, class, proper shipping name, packing group, marine pollutants, and other applicable information should be listed.

15. Regulatory information

On the label the obligatory symbols and codes prescribed as binding by the legislation ensuring the health protection and safety of a user and the environment should be placed.

16. Other information

This entry includes all other information recognized by a supplier as important for ensuing the health protection and safety of a user or for the environmental protection, for example a full list of R-phrases, recommended restrictions on use of the substance or preparation, additional instructions, sources of key data used to compile the chemical safety card.

In the case of a revised chemical safety card, a clear indication of the information which has been added, deleted or updated should be given.

9.2. Appendix 2 – Chemical Safety Card of Carbon Tetrachloride

(This card has been prepared by Polish institutions: Central Institute of Labour Protection (CILP) and National Research Institute (NRI))

CARBON TETRACHLORIDE CCl_4

1. IDENTIFICATON OF THE SUBSTANCE

Names and synonyms
Polish: czterochlorek węgla, czterochlorometan, tetrachlorometan, tetrachlorek węgla, tetra
English: carbontetrachloride, methanetetrachloride, perchloromethane, tetrachloromethane
German: Tetrachlorkohlenstoff, Kohlenstofftetrachlorid, Tetrachlormethan, Chlorkohlenstoff
French: tétrachlorure de carbone, tétrachlorométhane
Russian: четырёххлористый углерод, тетрахлорометан, перхлорометан, фреон-10
IUPAC name: tetrachloromethane
Chemical formula: CCl_4

2. COMPOSITION/INFORMATION ON COMPONENTS

Carbon tetrachloride – primary substance

Substance classification: T, N, Carcin. Cat. 3; R: 40-23/24/25-48/23-52-53-59; S: 1/2-23-36/37-45-59-61
ICSC Number: 0024
CAS Number: 56-23-5
UN Number: 1846
RTECS Number: FG4900000
EC Index Number: 602-008-00-5
EC (EINECS) Number: 200-262-8

3. HAZARD IDENTIFICATION

Toxic substance; dangerous for the environment. Inconclusive evidence of carcinogenic potency. Toxic by inhalation, in contact with skin and if swallowed. Causes danger of serious damage to health by prolongated exposure through inhalation. Harmful to aquatic organisms; may cause long-term adverse effects in the aquatic environment. Dangerous for the ozone layer.

4. FIRST-AID MEASURES

Obligatory medicaments: oxygen, liquid paraffin.

INTOXICATION BY INHALATION

Conscious

First-aid prior to medical attention

Take the intoxicated person out of the site of exposure. Keep the affected person at rest in any comfortable position. Administer oxygen for breathing preferably through a mask. Call a physician.

Medical attention

Headache and vertigo resulting from the exposure to this substance and consequently a risk of aggravation of the intoxication symptoms justify transportation of the exposed person to a hospital in an emergency ambulance under the surveillance of a physician.

Unconscious

First-aid prior to medical attention

Move the intoxicated person out of the site of exposure. Lay the affected person down in a stable side position. Remove from the buccal cavity the dentures and any other movable foreign bodies. Suck out, with a syringe through a catheter, the nasal and buccal discharge. If breathing has stopped, perform artificial respiration by a mouth-to-mouth method or using an AMBU respirator (an artificial-respiration device consisting of a bag that is squeezed by hand). Install a permanent intravenous drip (nurse). Immediately call a physician.

Medical attention

Sudden loss of consciousness may result from other conditions aside the intoxication by carbon tetrachloride. If breathing is disordered, perform incubation of the inhalation tract and artificial respiration using an AMBU respirator.

Control the heart beat (ECG). Do not apply adrenaline or other katechol amines. Transportation to a hospital in an ambulance with resuscitation equipment is strongly recommended.

SKIN IRRITATION

First-aid prior to medical attention

Remove the clothing, immediately wash the irritated skin with plenty of running tepid water (with soap if there are no dermal effects).

Medical attention

Larger areas of skin wetting justify transportation to a hospital because of the risk of developing the intoxication symptoms.

EYE IRRITATION

First-aid prior to medical attention

Rinse eyes with large amounts of cool water for approximately 15 minutes. The affected person may perform eye washing on own rules. Avoid strong stream of water because of the risk of mechanical damage to the cornea.

Medical attention

Ensure an urgent ophthalmological consultation.

INTOXICATION BY INGESTION

Conscious

First-aid prior to medical attention

Immediately after swallowing (within 5 minutes) the intoxicated person should induce vomiting. Later on do not provoke vomiting. Serve a 150 ml dose of liquid paraffin to drink. Do not serve milk, oils or alcohols. Call a physician.

Medical attention

Transportation to a hospital in an ambulance with resuscitation equipment because of the risk of aggravation of the intoxication symptoms is strongly recommended.

Unconscious

First-aid prior to medical attention

Proceed as in the case of intoxication by inhalation.

Medical attention

Proceed as in the case of intoxication by inhalation.

5. FIRE-FIGHTING MEASURES

HAZCHEM code: 2Z

Specific hazards

Toxic, inflammable liquid. The vapour is heavier than air, gathering close to the ground and in lower room spaces. The vapour displays a retarding effect on explosiveness of many mixtures of flammable gases or vapours with air. Containers exposed to fire or high temperature may explode.

General recommendations

Alert the neighbourhood about the accident. Remove from the endangered area all the persons not taking part in elimination of the accident effects. Call a national fire brigade and police.

Fire: inflammable liquid. Containers exposed to fire and high temperature spill with a cool water, if possible remove the containers from the endangered area. Do not let the sludges left after the fire extinguishing enter the sewage and water systems.

Special protective equipment

Wear the protective clothing made from coated materials and a self-contained breathing apparatus.

Attention: products of decomposition are acutely toxic.

6. MEASURES IN CASE OF ACCIDENTAL RELEASE

General recommendations

As specified in 5.

Leakage

Protect the interceptor sewers, avoid a direct contact with the released substance, remove the ignition sources, if possible eliminate the leakage (shut off the inflow of the liquid, seal, place the damaged packing in a protective packing), in case of a large leakage embank the site where the liquid is gathering, disperse the vapours with a sprayed stream of water, raise the collected liquid with a pump, cover the small amounts of spilled liquid with an inert absorbent, collect into a covered container, wash the contaminated area with water.

7. HANDLING AND STORING

Handling: while handling the substance do not eat, do not drink, avoid contact with the liquid, avoid the vapour inhalation, observe the rules of personal hygiene, use the personal protective equipment (as specified in 8), work in a well ventilated room, avoid the dangerous effect of high temperature and fire on the substance.

Storing: store in original, properly labeled and sealed packaging in a cool, dry, well ventilated storage room with an unpermeatable flat floor and the internal water system. Store away from sources of heat. Protect from the light effect.

8. CONTROL OF THE EXPOSURE AND PERSONAL PROTECTION

Requirements for ventilation

The necessary means of the exposure control are the workplace exhaust ventilation removing the vapours from their emission sites and the general ventilation of the workplace. Sucking holes of the workplace ventilation are at the working level or below. Exhausts of the general ventilation are in the upper part of the room or along the floor.

Polish occupational exposure limit values for carbon tetrachloride

HAC[a]	20 mg/m^3 [d]
HAMC[b]	100 mg/m^3 [d]
HATC[c]	undetermined

a) Highest Admissible Concentration.
b) Highest Admissible Momentary Concentration.
c) Highest Admissible Threshold Concentration.
d) Determined in the air at the workplace: in milligrams per cubic air at 20°C and 101.3 kPa.

The details on the determination of the above limit values can be found in Polish Norms (PN) related to the occupational safety and health:

PN-75/Z-04074 sheet 01: Clean air protection. Investigations of the carbon tetrachloride contents. Determination of the carbon tetrachloride concentration at the workplace by a colorimetric method.

PN-77/Z-04074 sheet 02: Clean air protection. Investigations of the carbon tetrachloride contents. Determination of the carbon tetrachloride concentration at the workplace by a gas chromatography method.

Requirements for the personal protective equipment

If the concentration of the substance is stable and known the selection of the personal protective equipment should be made taking into account the substance concentration found at a given workplace, the exposure time, the activities performed by a worker and the recommendations given by a producer of the personal protective equipment. Use the protective clothing made from coated materials (i.e. viton), safety gloves (for instance of vinyl polyalcohole), goggles protecting against splashes (preferably with a half mask), respiratory protection: a half mask or a full-face mask with an A-type filter (protecting against organic gases, vapours with bp. >65° C). In case of oxygen depletion (concentration below 17 Vol.-%) or if the concentration of the compound exceeds 1 Vol.-%, use an autonomic or stationary isolating equipment.

In a zone endangered by explosion, use the clothing, glasses and boots made from anti-electrostatic materials.

In an emergency situation or if the concentration of the substance at the workplace is unknown, use the personal protective equipment isolating the whole body (gas-tight protective suit with an equipment isolating the respiratory system).

9. PHYSICAL AND CHEMICAL PROPERTIES

Basic properties

Molecular mass	153.8
Physical state at 20°C	liquid
Colour	colourless
Odour	sweetish
Melting point	-22.95°C
Boiling point	76.8°C
Density at 20°C	1.59 g/cm³
Relative vapour density	5.3
Vapour pressure	
– at 20°C	119.4 hPa
– at 50°C	411.9 hPa
Saturated vapour content	
– at 20°C	757 g/m³
– at 30°C	1115 g/m³
Solubility in water at 20°C	0.08% weight
Solubility in other solvents	dissolves in ethyl alcohol, ethyl ether, chloroform, benzene, solvent naphtha

Other properties

Critical temperature	283.15°C
Critical pressure	4.55 MPa
Index of refraction at 20°C	1.4607
Viscosity at 20°C	0.969 mPa•s
Specific capacity at 25°C	0.865 J/(g•K)

Heat of evaporation	
– at bp.	194.94 J/g
– at 25°C	210.86 J/g
Heat of combustion	–1.01 kJ/g
Normal octanol/water partition coefficient as log Pow	2.64

10. STABILITY AND REACTIVITY

Stability: stable under normal conditions.
Decomposes if exposed to light at normal temperature
Thermal decomposition at temperature ca. 400°C is slow, but it is rapid at temperature 900÷1300°C.
Heating of the vapour/air mixture to 335°C in the presence of iron leads to formation of significant amounts of phosgene.

Conditions to avoid: high temperature, exposure to light, presence of water/moisture – at temperature ca. 250°C carbon dioxide and hydrogen chloride are formed, and also phosgene if small moisture contents are present.

Materials to avoid: dry CCl_4 does not react with majority of metals used for construction, it reacts slowly with copper and lead. In the presence of moisture the corroding effect on metals is observed.
CCl_4 reacts rapidly and explosively with aluminium dust, magnesium dust, zinc dust, beryllium dust, boron, sodium, potassium, sodium alloys, potassium alloys, trichloride of triethyldialuminium, calcium silicide, chlorine trifluoride, decaborane, ethylene, dinitrogen tetroxide, fluorine, borates, dimethylphosphamid in the presence of iron.

Dangerous combustion/decomposition products: on contact with hot surfaces or flames this substance decomposes forming acutely toxic products: chlorine, hydrogen chloride, phosgene.

11. TOXICOLOGICAL INFORMATION

Toxicity class
Toxic, carcinogenic (cat. 3) and dangerous for the environment according to the list of dangerous substances.
Not included in the lists of human carcinogens and probable human carcinogens (according to the regulation of the Polish Ministry of Health and Social Welfare, of 11 September 1996).
The substance is probably carcinogenic to humans according to IARC (group 2B).

Acute intoxication: lethal doses and toxic concentrations
Odour detection threshold – 64 mg/m^3
LD_{50} (rat, oral) – 2800 mg/kg
LD_{50} (rat, dermal) – 5070 mg/kg
LC_{50} (rat, inhalative) – 51200 mg/m^3 (4 h)
LD_{Lo} (man, oral) – 43 mg/kg
LC_{Lo} (man, inhalative) – 6400 mg/m^3

Depending on the incorporation pathway, the following experimental durations for the determination of the acute toxicity are used:
– oral: single application of the entire amount in the stomach, administered by gavage,
– dermal: single application of the entire amount of the substance to the skin, 24 h action duration,
– inhalative: exposure via breathed air for 4 h.
For a description of acute intoxication by substances, the following median lethal doses and concentrations are usually used:
– LD_{50} oral: dose at which half of the animals died by oral incorporation,
– LD_{50} dermal: dose at which half of the animals died by skin contact,
– LC_{50} inhalative: concentration at which half of the animals died by 4 h inhalative exposure.

The LD_{50} values are calculated by division of the median lethal amount of a substance on oral or dermal application by the body weight of the animals. For inhalative administration, on the other hand, the concentration of the substance in milligrams per cubic meter of breathed air after exposure for 4 h is used (LC_{50}).
The LD_{Lo} is the lowest dosage per unit of bodyweight (typically stated in milligrams per kilogram) of a substance known to have resulted in fatality in a particular animal species. When quoting an LD_{Lo}, the particular species and method of administration (e.g. ingested, inhaled, intravenous) are typically stated. Similarly, the LC_{Lo} (lethal concentration low) is the lowest dosage of the substance in milligrams per cubic meter of breathed air at which death occurred.

Toxic effect and other harmful biological effects on the human organism: toxic, depressive effect on the central nervous system, hepatotoxic and nephrotoxic, irritating.

Routes of exposure: by inhalation, through the skin, by ingestion.

Symptoms of acute intoxication: vapour in concentrations slightly above the admissible value causes headache and vertigo, nausea after few hours of exposure, at higher concentrations may cause irritation to mucous membranes, eyes, throat, headache and vertigo, balance disorder, nausea, vomiting, sleepiness, consciousness disorder.

Lethal concentration for humans (LC_{Lo}) is estimated at ca. 6 g/m³.
The direct effect of the acute intoxication – independent of the aggravating narcotic symptoms – is the liver impairment with a jaundice and the kidneys damage with a renal insufficiency. Contamination of the skin results in redness and pain. Wetting of large skin fragments constitutes a danger of intoxication with the symptoms characteristic for intoxication by inhalation. Contamination of the eyes with a liquid carbon tetrachloride causes pain, tear dripping, conjunctiva redness with a risk of conjunctiva impairment. Contamination by ingestion results in nausea, vomiting, a stomach-ache, diarrhea, necrosis of the liver with a jaundice and a hepatic coma, and in the kidneys damage with a renal insufficiency as well as the cardiovascular disorder. Death occurs after few days.

Lethal dose for humans is 3–30 ml.

Symptoms of chronic intoxication: headache and vertigo; chronic dermatitis.

12. ECOLOGICAL INFORMATION

Admissible/emergency levels of the substance in the air: undetermined
Admissible contamination of the inland surface waters: undetermined
Admissible contamination of the sewage discharged to the water and soil: undetermined

LC_{50} – *Lethal concentraton: concentration with death of half of the species*
EC_{50} – *Effective concentration: concentration with a specific effect on half of the investigated species*
IC_{50} – *Inhibitory concentration: concentration at which half the plants or animals show an inhibition effect, e.g. inhibition of growth*
LC_{100}, EC_{100} and IC_{100} *same as above but concern 100% population of the investigated species*

Toxic concentrations for animals and plants: Data for classification
Acute toxicity (LC_{50}/96 h) for fish *Lepomis macrochirus* – 125 mg/l
Acute toxicity (EC_{50}/48 h) for crustacea – no data
Inhibition of growth of algae (IC_{50}/72 h) – no data
Inhibition of growth of bacteria colonies – no data

Other data
Toxic concentration limit values for:
  – fish *Leuciscus idus melanotus* – 5 mg/l (LC_{50}/48 h)
  – crustacean *Daphnia magna* – 0.9 mg/l (EC_{50}/24 h)
  – bacteria: *Pseudomonas putida* – 30 mg/l
                    Microcystis aeruginosa – 105 mg/l
  – alga *Scenedesmus quadricauda* – 600 mg/l
  – protozoan *Entosiphon sulcatum* – 770 mg/l
Lethal concentration for:
  – fish *Leuciscus idus melanotus* – 13 mg/l (LC_{50}/48 h), 33 mg/l (LC_{100}/48 h)
  – crustacean *Daphnia magna* – 28 mg/l (EC_{50}/24 h), 159 mg/l (EC_{100}/24 h)
Concentration with no effect on the processes of biological treatment – 50 mg/l
Limit value for a harmful impact on anaerobic digestion – 180 mg/l

13. DISPOSAL CONSIDERATIONS

Waste classification: appropriate for the site of the waste generation on the basis of criteria included in legally binding regulations.
Waste treatment
Do not release into the water system. Do not dispose on the municipal dumps. Consider a possibility of reuse. Recovery or disposal of the waste product must be performed according to the legally binding regulations.
Recommended method of the waste neutralization: thermal or physicochemical treatment.
Small amounts of carbon tetrachloride can be hydrolyzed at elevated temperature using a 20% solution of sodium hydroxide.
Large amounts of carbon tetrachloride should be destroyed by incineration in a special furnace after mixing with combustible solvents.

Waste packaging treatment
The single-use packagings should be transferred to a legitimate waste recipient. Recycling or disposal must be carried out in accordance with the legally binding regulations. If it is necessary, the cleaned multiple-use packagings may be reused.

14. TRANSPORT INFORMATION

The substance is subject to the regulations on the transportation of dangerous goods.

UN material identification number	1846
Correct transport name	CARBON TETRACHLORIDE
Class	6.1
Classification code	T1
Packing group	II
Parking instruction	P001, IBC02
Hazard identification number	60
Parcel labels	UN 1846, No. 6.1
Transportation labels	
vehicles	
wagons	No. 6.1
tanks	60 / 1846 No. 6.1

15. REGULATORY INFORMATION
Packaging labels
Identification: carbon tetrachloride
EC (EINECS) number: 200-262-8 EC labelling

Danger symbols:

 Toxic (T) substance   Environmentally (N) dangerous substance

Phrases indicating special risks (R)

23/24/25 Toxic by inhalation, in contact with skin and if swallowed
40 Limited evidence of a carcinogenic effect
48/23 Toxic: danger of serious damage to health by prolonged exposure through inhalation
52/53 Harmful to aquatic organisms, may cause long-term adverse effects in the aquatic environment
59 Dangerous for the ozone layer

Phrases describing safety precautions concerning usage (S)

1/2 Keep locked up and keep out of the reach of children
23 Do not breathe vapour
36/37 Wear suitable protective clothing and suitable gloves
45 In case of accident or if you feel unwell, seek medical advice immediately (show label where possible)
59 Refer to manufacturer/supplier for information on recovery/recycling
61 Avoid release to the environment. Refer to special instructions or safety data sheet

National regulations
As specified in 6.

16. OTHER INFORMATION

Register number	0010
Card preparation date	15.10.1993
Card revision date	2004
Endorsed by	Programme Council

A similar but abbreviated version of the above chemical safety card prepared in Poland by CIWP and NRI can be found as an International Chemical Safety Card (ISCS) No. 0024 in the files stored by the Commission of the European Communities, National Institute of Occupational Safety and Health, and International Programme on Chemical Safety.

9.3. Appendix 3 – Phrases Indicating Special Risks and Describing Safety Precautions

INDICATION OF PARTICULAR RISKS AND THEIR CODES [R PHRASES]
(Journal of Laws of the Polish Republic 1997, No. 105, item 671, with amendments in Journal of Laws of the Polish Republic 1999, No. 26, item 261)

R1   Explosive when dry
R2   Risk of explosion by shock, friction, fire or other sources of ignition
R3   Extreme risk of explosion by shock, friction, fire or other sources of ignition
R4   Forms very sensitive explosive metallic compounds
R5   Heating may cause an explosion
R6   Explosive with or without contact with air
R7   May cause fire
R8   Contact with combustible material may cause fire
R9   Explosive when mixed with combustible material
R10 Flammable
R11 Highly flammable
R12 Extremely flammable
R14 Reacts violently with water
R15 Contact with water liberates extremely flammable gases
R16 Explosive when mixed with oxidizing substances
R17 Spontaneously flammable in air
R18 In use may form flammable/explosive vapor-air mixture
R19 May form explosive peroxides
R20 Harmful by inhalation
R21 Harmful in contact with skin
R22 Harmful if swallowed
R23 Toxic by inhalation
R24 Toxic in contact with skin
R25 Toxic if swallowed
R26 Very toxic by inhalation
R27 Very toxic in contact with skin
R28 Very toxic if swallowed
R29 Contact with water liberates toxic gas
R30 Can become highly flammable in use
R31 Contact with acids liberates toxic gas
R32 Contact with acids liberates very toxic gas
R33 Danger of cumulative effects
R34 Causes burns
R35 Causes severe burns
R36 Irritating to the eyes

R37 Irritating to the respiratory system
R38 Irritating to the skin
R39 Danger of very serious irreversible effects
R40 Limited evidence of a carcinogenic effect
R41 Risk of serious damage to eyes
R42 May cause sensitization by inhalation
R43 May cause sensitization by skin contact
R44 Risk of explosion if heated under confinement
R45 May cause cancer
R46 May cause heritable genetic damage
R48 Danger of serious damage to health by prolonged exposure
R49 May cause cancer by inhalation
R50 Very toxic to aquatic organisms
R51 Toxic to aquatic organisms
R52 Harmful to aquatic organisms
R53 May cause long-term adverse effects in the aquatic environment
R54 Toxic to flora
R55 Toxic to fauna
R56 Toxic to soil organisms
R57 Toxic to bees
R58 May cause long-term adverse effects in the environment
R59 Dangerous for the ozone layer
R60 May impair fertility
R61 May cause harm to the unborn child
R62 Possible risk of impaired fertility
R63 Possible risk of harm to the unborn child
R64 May cause harm to breast-fed babies
R65 Harmful: may cause lung damage if swallowed
R66 Repeated exposure may cause skin dryness or cracking
R67 Vapors may cause drowsiness and dizziness
R68 Possible risk of irreversible effects

Combined R phrases

R14/15	Reacts violently with water, liberating extremely flammable gases
R15/29	Contact with water liberates toxic, extremely flammable gas
R20/21	Harmful by inhalation and in contact with skin
R20/22	Harmful by inhalation and if swallowed
R20/21/22	Harmful by inhalation, in contact with skin and if swallowed
R21/22	Harmful in contact with skin and if swallowed
R23/24	Toxic by inhalation and in contact with skin
R23/25	Toxic by inhalation and if swallowed
R23/24/25	Toxic by inhalation, in contact with skin and if swallowed
R24/25	Toxic in contact with skin and if swallowed
R26/27	Very toxic by inhalation and in contact with skin
R26/28	Very toxic by inhalation and if swallowed

R26/27/28	Very toxic by inhalation, in contact with skin and if swallowed
R27/28	Very toxic in contact with skin and if swallowed
R36/37	Irritating to eyes and respiratory system
R36/38	Irritating to eyes and skin
R36/37/38	Irritating to eyes, respiratory system and skin
R37/38	Irritating to respiratory system and skin
R39/23	Toxic: danger of very serious irreversible effects through inhalation
R39/24	Toxic: danger of very serious irreversible effects in contact with skin
R39/25	Toxic: danger of very serious irreversible effects if swallowed
R39/23/24	Toxic: danger of very serious irreversible effects through inhalation in contact with skin
R39/23/25	Toxic: danger of very serious irreversible effects through inhalation and if swallowed
R39/24/25	Toxic: danger of very serious irreversible effects in contact with skin and if swallowed
R39/23/24/25	Toxic: danger of very serious irreversible effects through inhalation, in contact with skin and if swallowed
R39/26	Very toxic: danger of very serious irreversible effects through inhalation
R39/27	Very toxic: danger of very serious irreversible effects in contact with skin
R39/28	Very toxic: danger of very serious irreversible effects if swallowed
R39/26/27	Very toxic: danger of very serious irreversible effects through inhalation and in contact with skin
R39/26/28	Very toxic: danger of very serious irreversible effects through inhalation and if swallowed
R39/27/28	Very toxic: danger of very serious irreversible effects in contact with skin and if swallowed
R39/26/27/28	Very toxic: danger of very serious irreversible effects through inhalation, in contact with skin and if swallowed
R42/43	May cause sensitization by inhalation and skin contact
R48/20	Harmful: danger of serious damage to health by prolonged exposure through inhalation
R48/21	Harmful: danger of serious damage to health by prolonged exposure in contact with skin
R48/22	Harmful: danger of serious damage to health by prolonged exposure if swallowed
R48/20/21	Harmful: danger of serious damage to health by prolonged exposure through inhalation and in contact with skin
R48/20/22	Harmful: danger of serious damage to health by prolonged exposure through inhalation and if swallowed
R48/21/22	Harmful: danger of serious damage to health by prolonged exposure in contact with skin and if swallowed
R48/20/21/22	Harmful: danger of serious damage to health by prolonged exposure through inhalation, and in contact with skin and if swallowed
R48/23	Toxic: danger of serious damage to health by prolonged exposure through inhalation

R48/24	Toxic: danger of serious damage to health by prolonged exposure in contact with skin
R48/25	Toxic: danger of serious damage to health by prolonged exposure if swallowed
R48/23/24	Toxic: danger of serious damage to health by prolonged exposure through inhalation and in contact with skin
R48/23/25	Toxic: danger of serious damage to health by prolonged exposure through inhalation and if swallowed
R48/24/25	Toxic: danger of serious damage to health by prolonged exposure in contact with skin and if swallowed
R48/23/24/25	Toxic: danger of serious damage to health by prolonged exposure through inhalation, in contact with skin and if swallowed
R50/53	Very toxic to aquatic organisms, may cause long-term adverse effects in the aquatic environment
R51/53	Toxic to aquatic organisms, may cause long-term adverse effects in the aquatic environment
R52/53	Harmful to aquatic organisms, may cause long-term adverse effects in the aquatic environment
R68/20	Harmful: possible risk of irreversible effects through inhalation
R68/21	Harmful: possible risk of irreversible effects in contact with skin
R68/22	Harmful: possible risk of irreversible effects if swallowed
R68/20/21	Harmful: possible risk of irreversible effects through inhalation and in contact with skin
R68/20/22	Harmful: possible risk of irreversible effects through inhalation and if swallowed
R68/21/22	Harmful: possible risk of irreversible effects in contact with skin and if swallowed
R68/20/21/22	Harmful: possible risk of irreversible effects through inhalation, in contact with skin and if swallowed

ADDITIONAL PHRASES INDICATING PARTICULAR RISKS FOR PEOPLE AND ENVIRONMENT CREATED BY PESTICIDES, THEIR CODES AND APPLICATION CRITERIA

RSh1	Toxic in contact with eyes
RSh2	May cause hypersensitivity to light
RSh3	Contact with vapours causes burns to the skin and eyes, contact with liquid causes frostbites

PHRASES DESCRIBING SAFETY PRECAUTIONS CONCERNING USAGE OF DANGEROUS SUBSTANCES AND PREPARATIONS AND THEIR CODES [S PHRASES] (Journal of Laws of the Polish Republic 2002, No. 149, item 1173, with amendments in Journal of Laws of the Polish Republic 2003, No. 173, item 1679 and Journal of Laws of the Polish Republic 2004, No. 260, item 2595)

| S1 | Keep locked up |
| S2 | Keep out of the reach of children |

S3 Keep in a cool place
S4 Keep away from living quarters
S5 Keep contents under ... (appropriate liquid to be specified by the manufacturer)
S6 Keep under ... (inert gas to be specified by the manufacturer)
S7 Keep container tightly closed
S8 Keep container dry
S9 Keep container in a well-ventilated place
S12 Do not keep the container sealed
S14 Keep away from food, drink and animal feeding stuffs
S15 Keep away from heat
S16 Keep away from sources of ignition – No smoking
S17 Keep away from combustible material
S18 Handle and open container with care
S20 When using, do not eat or drink
S21 When using, do not smoke
S22 Do not breathe dust
S23 Do not breathe gas/fumes/vapour/spray (appropriate wording to be specified by the manufacturer)
S24 Avoid contact with skin
S25 Avoid contact with eyes
S26 In case of contact with eyes, rinse immediately with plenty of water and seek medical advice
S27 Take off immediately all contaminated clothing
S28 After contact with skin, wash immediately with plenty of ... (to be specified by the manufacturer)
S29 Do not empty into drains
S30 Never add water to this product
S33 Take precautionary measures against static discharges
S35 This material and its container must be disposed of in a safe way
S36 Wear suitable protective clothing
S37 Wear suitable gloves
S38 In case of insufficient ventilation, wear suitable respiratory equipment
S39 Wear eye/face protection
S40 To clean the floor and all objects contaminated by this material use ... (to be specified by the manufacturer)
S41 In case of fire and/or explosion do not breathe fumes
S42 During fumigation/spraying wear suitable respiratory equipment (appropriate wording to be specified)
S43 In case of fire, use ... (indicate in the space the precise type of fire-fighting equipment. If water increases the risk, add 'Never use water')
S45 In case of accident or if you feel unwell, seek medical advice immediately (show label where possible)
S46 If swallowed, seek medical advice immediately and show this container or label
S47 Keep at temperature not exceeding ... °C (to be specified by the manufacturer)
S48 Keep wetted with ... (appropriate material to be specified by the manufacturer)
S49 Keep only in the original container

S50   Do not mix with ... (to be specified by the manufacturer)
S51   Use only in well-ventilated areas
S52   Not recommended for interior use on large surface areas
S53   Avoid exposure – obtain special instruction before use
S56   Dispose of this material and its container to hazardous or special waste collection point
S57   Use appropriate container to avoid environmental contamination
S59   Refer to manufacturer/supplier for information on recovery/recycling
S60   This material and/or its container must be disposed of as hazardous waste
S61   Avoid release to the environment. Refer to special instructions or safety data sheet
S62   If swallowed, do not induce vomiting: seek medical advice immediately and show this container or label
S63   In case of accident by inhalation, remove casualty to fresh air and keep at rest
S64   If swallowed, rinse mouth with water (only if the person is conscious)

Combined S phrases

S1/2   Keep locked up and out of the reach of children
S3/7   Keep container tightly closed in a cool place
S3/9/14   Keep in a cool well-ventilated place away from (incompatible materials to be indicated by the manufacturer)
S3/9/14/49 Keep only in the original container in a cool well-ventilated place away from (incompatible materials to be indicated by the manufacturer)
S3/9/49   Keep only in the original container in a cool well-ventilated place
S3/14   Keep in a cool place away from ... (incompatible materials to be indicated by the manufacturer)
S7/8   Keep container tightly closed and dry
S7/9   Keep container tightly closed and in a well-ventilated place
S7/47   Keep container tightly closed and at a temperature not exceeding ...°C (to be specified by the manufacturer)
S20/21   When using, do not eat, drink or smoke
S24/25   Avoid contact with skin and eyes
S27/28   After contact with skin, take off immediately all contaminated clothing and wash immediately with plenty of ... (to be specified by the manufacturer)
S29/35   Do not empty into drains, dispose of this container and it's material in a safe way
S29/56   Do not empty into drains, dispose of this material and its container to hazardous or special waste-collection point
S36/37   Wear suitable protective clothing and gloves
S36/37/39   Wear suitable protective clothing, gloves and eye/face protection
S36/39   Wear suitable protective clothing and eye/face protection
S37/39   Wear suitable gloves and eye/face protection
S47/49   Keep only in the original container at temperature not exceeding ...°C (to be specified by the manufacturer)

9.4. Appendix 4 – Catalogue of Wastes Including Hazardous Wastes

(Journal of Laws of the Polish Republic, No 112, item 1206 of 27 September 2001)

01 **Wastes resulting from exploration, mining, quarrying, physical and chemical treatment of ores and other minerals**

01 01 **Wastes from mineral excavation**

01 01 01 Wastes from mineral metalliferous excavation (excluding 01 01 80)

01 01 02 Wastes from mineral non-metalliferous excavation

01 01 80 Rock wastes from copper, zinc and lead mining

01 03 **Wastes from physical and chemical processing of metalliferous minerals**

01 03 04* Acid-generating tailings from processing of sulphide ore

01 03 05* Other tailings containing dangerous substances (excluding 01 03 80)

01 03 06 Tailings other than those mentioned in 01 03 04, 01 03 05, 01 03 80 and 01 03 81

01 03 07* Other wastes containing dangerous substances from physical and chemical processing of metalliferous minerals

01 03 08 Dusty and powdery wastes other than those mentioned in 01 03 07

01 03 09 Red mud from alumina production other than the wastes mentioned in 01 03 07

01 03 80 *Wastes from enrichment of non-ferrous ores by floatation containing dangerous substances*

01 03 81 *Wastes from enrichment of non-ferrous ores by floatation other than those mentioned in 01 03 80*

01 03 99 Wastes not otherwise specified

01 04 **Wastes from physical and chemical processing of non-metalliferous minerals**

01 04 07* Wastes containing dangerous substances from physical and chemical processing of non-metalliferous minerals

01 04 08 Waste gravel and crushed rocks other than those mentioned in 01 04 07

01 04 09 Waste sand and clays

01 04 10 Dusty and powdery wastes other than those mentioned in 01 04 07

01 04 11 Wastes from potash and rock-salt processing other than those mentioned in 01 04 07

01 04 12 Tailings and other wastes from washing and cleaning of minerals other than those mentioned in 01 04 07 and 01 04 11

01 04 13 Wastes from stone cutting and sawing other than those mentioned in 01 04 07

01 04 80 *Wastes from enrichment of coal by floatation containing dangerous substances*

01 04 81 *Wastes from enrichment of coal by floatation other than those mentioned in 01 04 80*

01 04 82* *Wastes from enrichment of sulfur ores by floatation containing dangerous*
substances

01 04 83 *Wastes from enrichment of sulfur ores by floatation other than those men-*
tioned in 01 04 82

01 04 84* *Wastes from enrichment of phosphorus ores (phosphorites, apatites) by flo-*
atation containing dangerous substances

01 04 85 *Wastes from enrichment of phosphorus ores (phosphorites, apatites) by flo-*
atation other than those mentioned in 01 04 84

01 04 99 Wastes not otherwise specified

01 05   Drilling muds and other drilling wastes
01 05 04 Fresh-water drilling muds and wastes
01 05 05* Drillling muds and wastes containing oil
01 05 06* Drillling muds and wastes containing dangerous substances
01 05 07 Barite-containing drilling muds and wastes other than those mentioned in
01 05 05 and 01 05 06
01 05 08 Chloride-containing drilling muds and wastes other than those mentioned in
01 05 05 and 01 05 06
01 05 99 Wastes not otherwise specified

02   Wastes from agriculture, horticulture, aquaculture, forestry, hunting
and fishing, food preparation and processing
02 01   Wastes from agriculture, horticulture, aquaculture, forestry, hunting
and fishing
02 01 01 Sludges from washing and cleaning
02 01 02 Animal-tissue waste
02 01 03 Plant-tissue waste
02 01 04 Waste plastics (except packaging)
02 01 06 Animal faeces
02 01 07 Wastes from forestry
02 01 08* Agrochemical waste containing dangerous substances including pesticides
in toxicity class I and II (very toxic and toxic)
02 01 09 Agrochemical waste other than those mentioned in 02 01 08
02 01 10 Waste metal
02 01 80* *Dead and out of necessity slaughtered animals and animal-tissue waste dis-*
playing hazardous properties
02 01 81 *Dead animals and animal-tissue waste representing extremely hazardous*
materials other than those mentioned in 02 01 80
02 01 82 *Dead and out of necessity slaughtered animals*
02 01 83 *Wastes from horticulture*
02 01 99 Wastes not otherwise specified

02 02   Wastes from the preparation and processing of meat, fish and other fo-
ods of animal origin
02 02 01 Sludges from washing and cleaning
02 02 02 Animal-tissue waste
02 02 03 Materials unsuitable for consumption or processing
02 02 04 Sludges from on-site effluent treatment
02 02 80* *Animal-tissue waste displaying hazardous properties*

02 02 81 *Animal-tissue waste including wastes from the production of meat-bone feeding stuffs representing extremely hazardous materials other than those mentioned in   02 02 80*
02 02 82 *Wastes from the production of fish-meal other than those mentioned in 02 02 80*
02 02 99 Wastes not otherwise specified

02 03 **Wastes from the preparation and processing of edible products and substances, and vegetable wastes including wastes from fruit, vegetables, cereals, edible oils, cocoa, coffee, tea, and tobacco preparation and processing; yeast and yeast extract production; molasses preparation and fermentation (except 02 07)**
02 03 01 Sludges from washing, cleaning, peeling, centrifuging and separation
02 03 02 Wastes from preserving agents
02 03 03 Wastes from solvent extraction
02 03 04 Materials unsuitable for consumption and processing
02 03 05 Sludges from on-site effluent treatment
02 03 80 *Pomace, sludges and other wastes from processing of vegetable products (except 02 03 81)*
02 03 81 *Wastes form manufacture of vegetable feeding stuffs*
02 03 83 *Tobacco wastes*
02 03 99 Wastes not otherwise specified

02 04 **Wastes from sugar processing**
02 04 01 Sludges from cleaning and washing beet
02 04 02 Off-specification calcium carbonate and sugar chalk (defecation lime)
02 04 03 Sludges from on-site effluent treatment
02 04 80 *Beet pulp*
02 04 99 Wastes not otherwise specified

02 05 **Wastes from the diary products industry**
02 05 01 Materials unsuitable for consumption and processing
02 05 02 Sludges from on-site effluent treatment
02 05 80 *Waste whey*
02 06 99 Wastes not otherwise specified

02 06 **Wastes from the baking and confectionery industry**
02 06 01 Materials unsuitable for consumption and processing
02 06 02 Wastes from preserving agents
02 06 03 Sludges from on-site effluent treatment
02 06 80 *Edible fat unsuitable for use*
02 06 99 Wastes not otherwise specified

02 07 **Wastes from the production of alcoholic and non-alcoholic beverages (except coffee, tea and cocoa)**
02 07 01 Wastes from washing, cleaning and mechanical reduction of raw materials
02 07 02 Wastes from spirits distillation
02 07 03 Wastes from chemical treatment
02 07 04 Materials unsuitable for consumption and processing
02 07 05 Sludges from on-site effluent treatment
02 07 80 *Pomace, must and digested sludges, slops*
02 07 99 Wastes not otherwise specified

03   Wastes from wood processing and the production of panels and furnitu-re, pulp, paper and cardboard

03 01   Wastes from wood processing and the production of panels and furnitu-re

03 01 01   Waste bark and cork

03 01 04* Sawdust, shavings, cuttings, wood, particle board and veneer containing dangerous substances

03 01 05   Sawdust, shavings, cuttings, wood, particle board and veneer other than tho-se mentioned in 03 01 04

03 01 80 *Wastes from chemical wood processing containing dangerous substances*

03 01 81 *Wastes from chemical wood processing other than those mentioned in 03 01 80*

03 01 82 *Sludges from on-site effluent treatment*

03 01 99   Wastes not otherwise specified

03 02   Wastes from wood preservation

03 02 01* Non-halogenated organic wood preservatives and impregnants

03 02 02* Organochlorinated wood preservatives and impregnants

03 02 03* Organometallic wood preservatives and impregnants

03 02 04* Inorganic wood preservatives and impregnants

03 02 05* Other wood preservatives and impregnants containing dangerous substances

03 02 99   Wastes not otherwise specified

03 03   Wastes from pulp, paper abd cardboard production and processing

03 03 01   Waste bark and wood

03 03 02   Sludges (including green liquor sludge) from the production of cellulose ap-plying the sulfite process

03 03 05   De-inking sludges from paper recycling

03 03 07   Mechanically separated rejects from pulping of waste paper and cardboard

03 03 08   Wastes of sorting paper and cardboard destined for recycling

03 03 09   Lime mud waste

03 03 10   Fibre rejects, fibre-, filler- and coating sludges from mechanical separation

03 03 11   Sludges from on-site effluent treatment other than those mentioned in 03 03 10

03 03 80 *Sludges from chlorine and hypochlorite bleaching processes*

03 03 81 *Sludges from other bleaching processes*

03 03 99   Wastes not otherwise specified

04   Wastes from the leather, fur and textile industries

04 01   Wastes from the leather and fur industry

04 01 01   Wastes from fleshing (fleshings and lime splits)

04 01 02   Liming waste

04 01 03* Degreasing wastes containing solvents without a liquid phase

04 01 04   Tanning liquor containing chromium

04 01 05   Tanning liquor free of chromium

04 01 06   Sludges, in particular from on-site effluent treatment containing chromium

04 01 07   Sludges, in particular from on-site effluent treatment free of chromium

04 01 08   Waste tanned leather (shavings, cuttings, buffing dust) containing chromium

04 01 09   Wastes from dressing and finishing
04 01 99   Wastes not otherwise specified

0402   Wastes from the textile industry
04 02 09   Wastes from composite materials (impregnated textile, elastomer, plastomer)
04 02 10   Organic matter from natural products (e.g. grease, wax)
04 02 14*   Wastes from finishing containing organic solvents
04 02 15   Wastes from finishing other than those mentioned in 04 02 14
04 02 16*   Dyes and pigments containing dangerous substances
04 02 17   Dyes and pigments other than those mentioned in 04 02 16
04 02 19*   Wastes from on-site effluent treatment containing dangerous substances
04 02 20   Wastes from on-site effluent treatment other than those mentioned in
            04 02 19
04 02 21   Wastes from unprocessed textile fibres
04 02 22   Wastes from processed textile fibres
04 02 80   Wastes from wet treatment of textile products
04 02 99   Wastes not otherwise specified

05   Wastes from petroleum refining, natural gas purification and pyrolytic treatment of coal
0501   Wastes from petroleum processing (e.g. refining)
05 01 02*   Desalter sludges
05 01 03*   Tank bottom sludges
05 01 04*   Acid alkyl sludges
05 01 05*   Oil spills
05 01 06*   Oily sludges from maintenance operations of the plant or equipment
05 01 07*   Acid tars
05 01 08*   Other tars
05 01 09*   Sludges from on-site effluent treatment containing dangerous substances
05 01 10   Sludges from on-site effluent treatment other than those mentioned in
            05 01 09
05 01 11*   Wastes from cleaning of fuels with bases
05 01 12*   Oil containing acids
05 01 13   Boiler feedwater sludges
05 01 14   Wastes from cooling columns
05 01 15*   Spent natural filter materials (e.g. clays, muds)
05 01 16   Sulphur-containing wastes from petroleum desulphurisation
05 01 17   Bitumen
05 01 99   Wastes not otherwise specified

0506   Wastes from the pyrolytic treatment of coal
05 06 01*   Acid tars
05 06 03*   Other tars
05 06 04   Wastes from cooling columns
05 06 80 *Liquid wastes containing phenols*
05 06 99   Wastes not otherwise specified

0507   Wastes from natural gas purification and transportation
05 07 01*   Wastes containing mercury

05 07 02   Wastes containing sulphur
05 07 99   Wastes not otherwise specified

06   Wastes from the manufacture, formulation, supply and use of the products of inorganic chemical industry

06 01   Wastes from the manufacture, formulation, supply and use of inorganic acids

06 01 01* Sulphuric acid and sulphurous acid
06 01 02* Hydrochloric acid
06 01 03* Hydrofluoric acid
06 01 04* Phosphoric acid and phosphorous acid
06 01 05* Nitric acid and nitrous acid
06 01 06* Other acids
06 01 99   Wastes not otherwise specified

06 02   Wastes from the manufacture, formulation, supply and use of bases

06 02 01* Calcium hydroxide
06 02 03* Ammonium hydroxide
06 02 04* Sodium hydroxide and potassium hydroxide
06 02 05* Other bases
06 02 99   Wastes not otherwise specified

06 03   Wastes from the manufacture, formulation, supply and use of salts, their solutions and metallic oxides

06 03 11* Solid salts and solutions containing cyanides
06 03 13* Solid salts and solutions containing heavy metals
06 03 14   Solid salts and solutions other than those mentioned in 06 03 11 and 06 03 13
06 03 15* Metallic oxides containing heavy metals
06 03 16   Metallic oxides other than those mentioned in 06 03 15
06 03 99   Wastes not otherwise specified

06 04   Metal-containing wastes other than those mentioned in 06 03

06 04 03* Wastes containing arsenic
06 04 04* Wastes containing mercury
06 04 05* Wastes containing other heavy metals
06 04 99   Wastes not otherwise specified

06 05   Sludges from on-site effluent treatment

06 05 02* Sludges from on-site effluent treatment containing dangerous substances
06 05 03   Sludges from on-site effluent treatment other than those mentioned in 06 05 02

06 06   Wastes from the manufacture, formulation, supply and use of sulphur chemicals, sulphur chemical processes and desulphurisation processes

06 06 02* Wastes containing dangerous sulphides
06 06 03   Wastes containing sulphides other than those mentioned in 06 06 02
06 06 99   Wastes not otherwise specified

06 07   Wastes from the manufacture, formulation, supply and use of halogen and halogen chemical processes

06 07 01* Wastes containing asbestos from electrolysis

06 07 02* Activated carbon from chlorine production
06 07 03* Barium sulphate sludge containing mercury
06 07 04* Solutions and acids (e.g. sulphuric acid)
06 07 99 Wastes not otherwise specified

06 08 Wastes from the manufacture, formulation, supply and use of silicon and silicon derivatives
06 08 02* Wastes containing dangerous silicones
06 08 99 Wastes not otherwise specified

06 09 Wastes from the manufacture, formulation, supply and use phosphorous chemicals and phosphorous chemical processes
06 09 02 Phosphorous slag
06 09 03* Calcium-based reaction wastes containing or contaminated with dangerous substances
06 09 04 Calcium-based reaction wastes other than those mentioned in 06 09 03 and 06 09 80
06 09 80 Phosphogypsum
06 09 81 Phosphogypsum mixed with slag, bottom ash and boiler dust (except boiler dust mentioned in 10 01 04)
06 09 99 Wastes not otherwise specified

06 10 Wastes from the manufacture, formulation, supply and use of nitrogen chemicals, nitrogen chemical processes and fertiliser manufacture
06 10 02* Wastes containing dangerous substances
06 10 99 Wastes not otherwise specified

06 11 Wastes from the manufucture of inorganic pigments and opacificiers
06 11 01 Calcium-based reaction wastes from titanium dioxide production
06 11 80 Wastes from the manufacture of zirconium compounds
06 11 81 Wastes from th emanufacture of chromium compounds
06 11 82 Wastes from the manufacture of cobalt compounds
06 11 83 Waste iron sulphate
06 11 99 Wastes not otherwise specified

06 13 Wastes from inorganic chemical processes not otherwise specified
06 13 01 Inorganic plant protection products (e.g. pesticides), wood-preserving agents and other biocides
06 13 02* Spent activated carbon (except 06 07 02)
06 13 03 Carbon black
06 13 04* Wastes from asbestos processing
06 13 05* Soot containing or contaminated with dangerous substances
06 13 99 Wastes not otherwise specified

07 Wastes from the manufacture, formulation, supply and use of the products of organic chemical industry
07 01 Wastes from the manufacture, formulation, supply and use of the basic products of organic chemical industry
07 01 01* Aqueous washing liquids and mother liquors
07 01 03* Organic halogenated solvents, washing liquids and mother liquors
07 01 04* Other organic solvents, washing liquids and mother liquors

07 01 07* Halogenated still bottoms and reaction residues
07 01 08* Other still bottoms and reaction residues
07 01 09* Halogenated filter cakes and spent sorbents
07 01 10* Other filter cakes and spent sorbents
07 01 11* Sludges from on-site effluent treatment containing dangerous substances
07 01 12 Sludges from on-site effluent treatment other than those mentioned in 07 01 11
07 01 80 Carbide residue free of dangerous substances (other than those mentioned in 07 01 08)
07 01 99 Wastes not otherwise specified

07 02 **Wastes from the manufacture, formulation, supply and use of plastics, synthetic rubberand man-made fibres**
07 02 01* Aqueous washing liquids and mother liquors
07 02 03* Organic halogenated solvents, washing liquids and mother liquors
07 02 04* Other organic solvents, washing liquids and mother liquors
07 02 07* Halogenated still bottoms and reaction residues
07 02 08* Other still bottoms and reaction residues
07 02 09* Halogenated filter cakes and spent sorbents
07 02 10* Other filter cakes and spent sorbents
07 02 11* Sludges from on-site effluent treatment containing dangerous substances
07 02 12 Sludges from on-site effluent treatment other than those mentioned in 07 02 11
07 02 13 Waste plastic
07 02 14* Wastes from additives containing dangerous substances
07 02 15 Wastes from additives other than those mentioned in 07 02 14
07 02 16* Wastes containing dangerous silicones
07 02 17 Wastes containing silicones other than those mentioned in 07 02 16
07 02 80 Wastes from rubber industry and rubber manufacture
07 02 99 Wastes not otherwise specified

07 03 **Wastes from the manufacture, formulation, supply and use of organic dyes and pigments (except 06 11)**
07 03 01* Aqueous washing liquids and mother liquors
07 03 03* Organic halogenated solvents, washing liquids and mother liquors
07 03 04* Other organic solvents, washing liquids and mother liquors
07 03 07* Halogenated still bottoms and reaction residues
07 03 08* Other still bottoms and reaction residues
07 03 09* Halogenated filter cakes and spent sorbents
07 03 10* Other filter cakes and spent sorbents
07 03 11* Sludges from on-site effluent treatment containing dangerous substances
07 03 12 Sludges from on-site effluent treatment other than those mentioned in 07 03 11
07 03 99 Wastes not otherwise specified

07 04 **Wastes from the manufacture, formulation, supply and use of organic plant protection products (except 02 01 08 and 02 01 09), wood preserving agents (except 03 02) and other biocides**
07 04 01* Aqueous washing liquids and mother liquors
07 04 03* Organic halogenated solvents, washing liquids and mother liquors
07 04 04* Other organic solvents, washing liquids and mother liquors

07 04 07* Halogenated still bottoms and reaction residues
07 04 08* Other still bottoms and reaction residues
07 04 09* Halogenated filter cakes and spent sorbents
07 04 10* Other filter cakes and spent sorbents
07 04 11* Sludges from on-site effluent treatment containing dangerous substances
07 04 12 Sludges from on-site effluent treatment other than those mentioned in 07 04 11
07 04 13* Solid wastes containing dangerous substances
07 04 80 Outdated pesticides in toxicity class I and II (very toxic and toxic)*
07 04 81 Outdated pesticides other than those mentioned in 07 04 80
07 04 99 Wastes not otherwise specified

07 05 **Wastes from the manufacture, formulation, supply and use of pharmaceuticals**
07 05 01* Aqueous washing liquids and mother liquors
07 05 03* Organic halogenated solvents, washing liquids and mother liquors
07 05 04* Other organic solvents, washing liquids and mother liquors
07 05 07* Halogenated still bottoms and reaction residues
07 05 08* Other still bottoms and reaction residues
07 05 09* Halogenated filter cakes and spent sorbents
07 05 10* Other filter cakes and spent sorbents
07 05 11* Sludges from on-site effluent treatment containing dangerous substances
07 05 12 Sludges from on-site effluent treatment other than those mentioned in 07 05 11
07 05 13* Solid wastes containing dangerous substances
07 05 14 Solid wastes other than those mentioned in 07 05 13
07 05 80 Liquid wastes containing dangerous substances*
07 05 81 Liquid wastes other than those mentioned in 07 05 80
07 05 99 Wastes not otherwise specified

07 06 **Wastes from the manufacture, formulation, supply and use of fats, grease, soaps, detergents, disinfectants and cosmetics**
07 06 01* Aqueous washing liquids and mother liquors
07 06 03* Organic halogenated solvents, washing liquids and mother liquors
07 06 04* Other organic solvents, washing liquids and mother liquors
07 06 07* Halogenated still bottoms and reaction residues
07 06 08* Other still bottoms and reaction residues
07 06 09* Halogenated filter cakes and spent sorbents
07 06 10* Other filter cakes and spent sorbents
07 06 11* Sludges from on-site effluent treatment containing dangerous substances
07 06 12 Sludges from on-site effluent treatment other than those mentioned in 07 06 11
07 06 80 Fuller's earth from oil refining
07 06 81 Returns of cosmetics and testers
07 06 99 Wastes not otherwise specified

07 07 **Wastes from the manufacture, formulation, supply and use of chemical products not otherwise specified**
07 07 01* Aqueous washing liquids and mother liquors
07 07 03* Organic halogenated solvents, washing liquids and mother liquors
07 07 04* Other organic solvents, washing liquids and mother liquors
07 07 07* Halogenated still bottoms and reaction residues

07 07 08* Other still bottoms and reaction residues
07 07 09* Halogenated filter cakes and spent sorbents
07 07 10* Other filter cakes and spent sorbents
07 07 11* Sludges from on-site effluent treatment containing dangerous substances
07 07 12 Sludges from on-site effluent treatment other than those mentioned in 07 07 11
07 07 99 Wastes not otherwise specified

08 **Wastes from the manufacture, formulation, supply and use of coatings (paints, varnishes and vitreous enamels), fillers, adhesives, sealants and printing inks**

08 01 **Wastes from the manufacture, formulation, supply and use and removal of paint and varnish**

08 01 11* Waste paint and varnish containing organic solvents and other dangerous substances
08 01 12 Waste paint and varnish other than those mentioned in 08 01 11
08 01 13* Sludges from removal of paint and varnish containing organic solvents and other dangerous substances
08 01 14 Sludges from removal of paint and varnish other than those mentioned in 08 01 13
08 01 15* Aqueous sludges containing paint and varnish containing organic solvents or other dangerous substances
08 01 16 Aqueous sludges containing paint and varnish other than those mentioned in 08 01 15
08 01 17* Wastes from removal of paint and varnish removal containing organic solvents or other dangerous substances
08 01 18 Wastes from removal of paint and varnish removal other than those mentioned in 08 01 17
08 01 19* Aqueous suspensions containing paint and varnish containing organic solvents or other dangerous substances
08 01 20 Aqueous suspensions containing paint and varnish other thean those mentioned in 08 01 19
08 01 21* Waste paint or varnish remover
08 01 99 Wastes not otherwise specified

08 02 **Wastes from the manufacture, formulation, supply and use of other coatings (including ceramic materials)**

08 02 01 Waste coating powders
08 02 02 Aqueous sludges containing ceramic materials
08 02 03 Aqueous suspensions containing ceramic materials
08 02 99 Wastes not otherwise specified

08 03 **Wastes from the manufacture, formulation, supply and use of printing inks**

08 03 07 Aqueous sludges containing printing ink
08 03 08 Aqueous liquid waste containing printing ink
08 03 12* Waste printing ink containing dangerous substances
08 03 13 Waste printing ink other than those mentioned in 08 03 12
08 03 14* Printing ink sludges containing dangerous substances
08 03 15 Printing ink sludges other than those mentioned in 08 03 14

08 03 16* Waste etching solutions

08 03 17* Waste printing toner containing dangerous substances

08 03 18 Waste printing toner other than those mentioned in 08 03 17

08 03 19* Disperse oil containing dangerous substances

08 03 80 Disperse oil other than those mentioned in 08 03 19

08 03 99 Wastes not otherwise specified

08 04 **Wastes from the manufacture, formulation, supply and use of adhesives and sealants (including water-proofing products)**

08 04 09* Waste adhesives and sealants containing organic solvents or other dangerous substances

08 04 10 Waste adhesives and sealants other than those mentioned in 08 04 09

08 04 11* Adhesive and sealant sludges containing organic solvents or other dangerous substances

08 04 12 Adhesive and sealant sludges other than those mentioned in 08 04 11

08 04 13* Aqueous sludges containing adhesives or sealants containing organic solvents or other dangerous substances

08 04 14 Aqueous sludges containing adhesives or sealants other than those mentioned in 08 04 13

08 04 15* Aqueous liquid waste containing adhesives or sealants containing organic solvents or other dangerous substances

08 04 16 Aqueous liquid waste containing adhesives or sealants other than those mentioned in 08 04 15

08 04 17* Rosin oil

08 04 99 Wastes not otherwise specified

08 05 **Wastes not otherwise specified in 08**

08 05 01* Waste isocyanates

09 **Wastes from the photographic industry and photographic services**

09 01 **Wastes from the photographic industry and photographic services**

09 01 01* Water-based developer and activator solutions

09 01 02* Water-based offset plate developer solutions

09 01 03* Solvent-based developer solutions

09 01 04* Fixer solutions

09 01 05* Bleach solutions and bleach fixer solutions

09 01 06* Wastes containing silver from on-site effluent treatment

09 01 07 Photographic film and paper containing silver or silver compounds

09 01 08 Photographic film and paper free of silver

09 01 10 Single-use cameras without batteries

09 01 11* Single-use cameras containing batteries mentioned in 16 06 01, 16 06 02 or 16 06 03

09 01 12 Single-use cameras containing batteries other than those mentioned in 09 01 11

09 01 13* Aqueous liquid waste from on-site reclamation of silver other than those mentioned in 09 01 06

09 01 80 Outdated photographic reagents*

09 01 99 Wastes not otherwise specified

10   **Wastes from thermal processes**

10 01   **Wastes from power stations and other combustion plants (except 19)**

10 01 01   Bottom ash, slag and boiler dust (excluding boiler dust mentioned in 10 01 04)

10 01 02   Coal fly ash

10 01 03   Fly ash from peat and untreated wood

10 01 04* Oil fly ash and boiler dust

10 01 05   Calcium-based reaction wastes from flue-gas desulphurisation in solid form

10 01 07   Calcium-based reaction wastes from flue-gas desulphurisation in sludge form

10 01 09* Sulphuric acid

10 01 13* Fly ash from emulsified hydrocarbons used as fuel

10 01 14* Bottom ash, slag and boiler dust from co-incineration containing dangerous substances

10 01 15   Bottom ash, slag and boiler dust from co-incineration other than those mentioned in 10 01 14

10 01 16* Fly ash from co-incineration containing dangerous substances

10 01 17   Fly ash from co-incineration other than those mentioned in 10 01 16

10 01 18* Wastes from flue-gas treatment containing dangerous substances

10 01 19   Wastes from flue-gas treatment other than those mentioned in 10 01 05, 10 01 07 and 10 01 18

10 01 20   Sludges from on-site effluent treatment containing dangerous substances

10 01 21   Sludges from on-site effluent treatment other than those mentioned in 10 01 20

10 01 22* Aqueous sludges from boiler cleansing containing dangerous substances

10 01 23   Aqueous sludges from boiler cleansing other than those mentioned in 10 01 22

10 01 24   Sands from fluidised beds (except 10 01 82)

10 01 25   Wastes from fuel storage and preparation of coal-fired power plants

10 01 26   Wastes from cooling-water treatment

10 01 80   Bottom ash and slag mixtures from slag-tap furnace

10 01 81   Microspheres of fly ash

10 01 82   Mixtures of fly ash and solid waste from calcium-based reaction of flue-gas desulphurisation (dry and semi-dry methods of combustion gas desulphurisation and fluidised combustion)

10 01 99   Wastes not otherwise specified

10 02   **Wastes from the iron and steel industry**

10 02 01   Slag from smelting processes (blast-furnace and steelmaking processes)

10 02 02   Unprocessed slag from other processes

10 02 07* Solid wastes from flue-gas treatment containing dangerous substances

10 02 08   Solid wastes from flue-gas treatment other than those mentioned in 10 02 07

10 02 10   Mill scales

10 02 11* Wastes from cooling-water treatment containing oil

10 02 12   Wastes from cooling-water treatment other than those mentioned in 10 02 11

10 02 13* Sludges and filter cakes from flue-gas treatment containing dangerous substances

10 02 14 Sludges and filter cakes from flue-gas treatment other than those mentioned in 10 02 13

10 02 15 Other sludges and filter cakes

10 02 80 Skimmings from iron metallurgy

10 02 81 Waste iron sulphate

10 02 99 Wastes not otherwise specified

10 03 **Wastes from aluminium metallurgy**

10 03 02 Anode scraps

10 03 04* Primary production slags

10 03 05 Waste alumina

10 03 08* Salt slags from secondary production

10 03 09* Black drosses from secondary production

10 03 15* Skimmings that are flammable or emit, upon contact with water, flammable gases in dangerous quantities

10 03 16 Skimmings other than those mentioned in 10 03 15

10 03 17* Tar-containing wastes from anode manufacture

10 03 18 Carbon-containing wastes from anode manufacture other than those mentioned in 10 03 07

10 03 19* Flue-gas dust containing dangerous substances

10 03 20 Flue-gas dust other than those mentioned in 10 03 19

10 03 21* Other particulates and dust (including ball-mill dust) containing dangerous substances

10 03 22 Other particulates and dust (including ball-mill dust) other than those mentioned in 10 03 21

10 03 23* Solid wastes from flue-gas treatment containing dangerous substances

10 03 24 Solid wastes from flue-gas treatment other than those mentioned in 10 03 23

10 03 25* Sludges and filter cakes from flue-gas treatment containing dangerous substances

10 03 26 Sludges and filter cakes from flue-gas treatment other than those mentioned in 10 03 25

10 03 27* Wastes from cooling-water treatment containing oil

10 03 28 Wastes from cooling-water treatment other than those mentioned in 10 03 27

10 03 29* Wastes from treatment of salt slags and black drosses containing dangerous substances

10 03 30 Wastes from treatment of salt slags and black drosses other than those mentioned in 10 03 29

10 03 99 Wastes not otherwise specified

10 04 **Wastes from lead metallurgy**

10 04 01* Slags from primary and secondary production

10 04 02* Skimmings from primary and secondary production

10 04 03* Lime containing arsenic compounds (calcium arsenate)

10 04 04* Flue-gas dust

10 04 05* Other particulates and dust

10 04 06* Solid wastes from flue-gas treatment

10 04 07* Sludges and filter cakes from flue-gas treatment

10 04 09* Wastes from cooling-water treatment containing oil

10 04 10   Wastes from cooling-water treatment other than those mentioned in 10 04 09
10 04 99   Wastes not otherwise specified

10 05   Wastes from zinc metallurgy
10 05 01   Slags from primary and secondary production (except 10 05 80)
10 05 03* Flue-gas dust
10 05 04   Other particulates and dust
10 05 05* Solid wastes from flue-gas treatment
10 05 06* Sludges and filter cakes from flue-gas treatment
10 05 08* Wastes from cooling-water treatment containing oil
10 05 09   Wastes from cooling-water treatment other than those mentioned in 10 05 08
10 05 10* Skimmings that are flammable or emit, upon contact with water, flammable gases in dangerous quantities
10 05 11   Skimmings other than those mentioned in 10 05 10
10 05 80   Granular slag from shaft furnaces and slag from rotary kilns
10 05 99   Wastes not otherwise specified

10 06   Wastes from copper metallurgy
10 06 01   Slags from primary and secondary production
10 06 02   Skimmings from primary and secondary production
10 06 03* Flue-gas dust
10 06 04   Other particulates and dust
10 06 06* Solid wastes from flue-gas treatment
10 06 07* Sludges and filter cakes from flue-gas treatment
10 06 09* Wastes from cooling-water treatment containing oil
10 06 10   Wastes from cooling-water treatment other than those mentioned in 10 06 09
10 06 80   Shaft furnace slag and granular slag
10 06 99   Wastes not otherwise specified

10 07   Wastes from silver, gold and platinum metallurgy
10 07 01   Slags from primary and secondary production
10 07 02   Skimmings from primary and secondary production
10 07 03   Solid wastes from flue-gas treatment
10 07 04   Other particulates and dust
10 07 05   Sludges and filter cakes from flue-gas treatment
10 07 07* Wastes from cooling-water treatment containing oil
10 07 08   Wastes from cooling-water treatment other than those mentioned in 10 07 07
10 07 99   Wastes not otherwise specified

10 08   Wastes from other non-ferrous metallurgies
10 08 04   Particulates and dust
10 08 08* Salt slag from primary and secondary production
10 08 09   Other slags
10 08 10* Skimmings that are flammable or emit, upon contact with water, flammable gases in dangerous quantities
10 08 11   Skimmings other than those mentioned in 10 08 10
10 08 12* Tar-containing wastes from anode manufacture
10 08 13   Carbon-containing wastes from anode manufacture other than those mentioned in 10 08 12

10 08 14   Anode scrap
10 08 15*   Flue-gas dust containing dangerous substances
10 08 16   Flue-gas dust other than those mentioned in 10 08 15
10 08 17*   Sludges and filter cakes from flue-gas treatment containing dangerous sub-
            stances
10 08 18   Sludges and filter cakes from flue-gas treatment other than those mentioned
            in 10 08 17
10 08 19*   Wastes from cooling-water treatment containing oil
10 08 20   Wastes from cooling-water treatment other than mentioned in 10 08 19
10 08 99   Wastes not otherwise specified

10 09   Wastes from casting of ferrous pieces
10 09 03   Furnace slag
10 09 05*   Casting cores and moulds which have not undergone pouring containing
            dangerous substances
10 09 06   Casting cores and moulds which have not undergone pouring other than tho-
            se mentioned in 10 09 05
10 09 07*   Casting cores and moulds which have undergone pouring containing dange-
            rous substances
10 09 08   Casting cores and moulds which have undergone pouring other than those
            mentioned in 10 09 07
10 09 09*   Flue-gas dust containing dangerous substances
10 09 10   Flue-gas dust other than those mentioned in 10 09 09
10 09 11*   Other particulates containing dangerous substances
10 09 12   Other particulates than those mentioned in 10 09 11
10 09 13*   Waste binders containing dangerous substances
10 09 14   Waste binders other than those mentioned in 10 09 13
10 09 15*   Waste crack-indicating agent containing dangerous substances
10 09 16   Waste crack-indicating agent other than those mentioned in 10 09 15
10 09 80   Waster cast-iron products
10 09 99   Wastes not otherwise specified

10 10   Wastes from casting non-ferrous pieces
10 10 03   Casting skimmings and furnace slag
10 10 05*   Casting cores and moulds which have not undergone pouring containing
            dangerous substances
10 10 07*   Casting cores and moulds which have undergone pouring containing dange-
            rous substances
10 10 08   Casting cores and moulds which have undergone pouring other than those
            mentioned in 10 10 07
10 10 09*   Flue-gas dust containing dangerous substances
10 10 10   Flue-gas dust other than those mentioned in 10 10 09
10 10 11*   Other particulates containing dangerous substances
10 10 12   Other particulates than those mentioned in 10 10 11
10 10 13*   Waste binders containing dangerous substances
10 10 14   Waste binders other than those mentioned in 10 10 13
10 10 15*   Waste crack-indicating agent containing dangerous substances
10 10 16   Waste crack-indicating agent other than those mentioned in 10 10 15

10 10 99   Wastes not otherwise specified

10 11   **Wastes from manufacture of glass and glass products**

10 11 03   Waste glass-based fibrous materials

10 11 05   Particulates and dust

10 11 09* Waste preparation mixture before thermal processing containing dangerous
substances

10 11 10   Waste preparation mixture before thermal processing other than those mentioned in 10 11 09

10 11 11* Waste glass in small particles and glass powder containing heavy metals
(e.g. from electron ray tubes)

10 11 12   Waste glass other than those mentioned in 10 11 11

10 11 13* Glass-polishing and glass-grinding sludge containing dangerous substances

10 11 14   Glass-polishing and glass-grinding sludge other than those mentioned in
10 11 13

10 11 15* Solid wastes from flue-gas treatment containing dangerous substances

10 11 16   Solid wastes from flue-gas treatment other than those mentioned in 10 11 15

10 11 17* Sludges and filter cakes from flue-gas treatment containing dangerous substances

10 11 18   Sludges and filter cakes from flue-gas treatment other than those mentioned
in 10 11 17

10 11 19* Solid wastes from on-site effluent treatment containing dangerous substances

10 11 20   Solid wastes from on-site effluent treatment other than those mentioned in
10 11 19

10 11 80   *Fluorosilicate sludges*

10 11 81* *Wastes containing asbestos*

10 11 99   Wastes not otherwise specified

10 12   **Wastes from manufacture of ceramic sanitaryware and whiteware, and
refractory ceramics (ceramic goods, bricks, tiles and construction products)**

10 12 01   Waste preparation mixture before thermal processing

10 12 03   Particulates and dust

10 12 05   Sludges and filter cakes from gas treatment

19 12 06   Discarded molds

10 12 08   Waste ceramics, bricks, tiles and construction products (after thermal processing)

10 12 09* Solid wastes from flue-gas treatment containing dangerous substances

10 12 10   Solid wastes from flue-gas treatment other than those mentioned in 10 12 09

10 12 11* Wastes from glazing containing heavy metals

10 12 12   Wastes from glazing other than those mentioned in 10 12 11

10 12 13   Sludges from on-site effluent treatment

10 12 99   Wastes not otherwise specified

10 13   **Wastes from manufacture of mineral binders (including cement, lime
and plaster) and products made from them**

10 13 01   Waste preparation mixture before thermal processing

10 13 04   Wastes from calcination and hydration of lime

10 13 06   Particulates and dust (except 10 13 12 and 10 13 13)

10 13 07   Sludges and filter cakes from flue-gas treatment
10 13 09* Wastes from asbestos-cement manufacture containing asbestos
10 13 10   Wastes from asbestos-cement manufacture other than those mentioned in
           10 13 09
10 13 11   Wastes from cement-based composite materials other than those mentioned
           in 10 13 09 and 10 13 10
10 13 12* Solid wastes from flue-gas treatment containing dangerous substances
10 13 13   Solid wastes from flue-gas treatment other than those mentioned in 10 13 12
10 13 14   Waste concrete and concrete sludge
10 13 80   Wastes from cement production
10 13 81   Wastes from plaster production
10 13 82   Waster products
10 13 99   Wastes not otherwise specified

10 14   Waste from crematoria
10 14 01* Waste from flue-gas cleaning containing mercury

10 80   Wastes from ferroalloy production
10 80 01   Slag from ferrosilicon production
10 80 02   Dust from ferrosilicon production
10 80 03   Slag from ferrochromium production
10 80 04   Dust from ferrochromium production
10 80 05   Slag from ferromanganese production
10 80 06   Dust from ferromanganese production
10 80 99   Wastes not otherwise specified

**11   Wastes from chemical surface treatment and coating of metals and
           other materials; non-ferrous hydrometallurgy**
**11 01   Wastes from chemical surface treatment and coating of metals and
           other materials (e.g. galvanic processes, zinc coating process, pickling
           processes, phosphating, alkaline degreasing, anodising)**
11 01 05* Pickling acids
11 01 06* Wastes containing acids other than those mentioned in 11 01 05
11 01 07* Pickling bases
11 01 08* Phosphatising sludges
11 01 09* Sludges and filter cakes containing dangerous substances
11 01 10 * Sludges and filter cakes other than those mentioned in 11 01 09
11 01 11* Aqueous rinsing liquids containing dangerous substances
11 01 12   Aqueous rinsing liquids other than those mentioned in 11 01 11
11 01 13* Degreasing wastes containing dangerous substances
11 01 14   Degreasing wastes other than those mentioned in 11 01 13
11 01 15* Eluate and sludges from membrane systems or ion exchange systems con-
           taining dangerous substances
11 01 16* Saturated or spent ion exchange resins
11 01 98 Other wastes containing dangerous substances*
11 01 99   Wastes not otherwise specified

11 02   Wastes and sludges from non-ferrous hydrometallurgical processes
11 02 02* Sludges from zinc hydrometallurgy (including jarosite, goethite)

11 02 03   Wastes from the production of anodes for aqueous electrolytical processes
11 02 05* Wastes from copper hydrometallurgical processes containing dangerous substances
11 02 06   Wastes from copper hydrometallurgical processes other than those mentioned in 11 02 05
11 02 07* Other wastes containing dangerous substances
11 02 99   Wastes not otherwise mentioned

11 03   Sludges and solid wastes from tempering processes
11 03 01* Wastes containing cyanide
11 03 02* Other wastes

11 05   Wastes from hot galvanising processes
11 05 01   Hard zinc
11 05 02   Zinc ash
11 05 03* Solid wastes from flue-gas treatment
11 05 04* Spent flux
11 05 99   Wastes not otherwise mentioned

12   Wastes from shaping and physical and mechanical surface treatment of metals and plastics
12 01   Wastes from shaping and physical and mechanical surface treatment of metals and plastics
12 01 01   Ferrous metal filings and turnings
12 01 02   Ferrous metal dust and particles
12 01 03   Non-ferrous metal filings and turnings
12 01 04   Non-ferrous metal dust and particles
12 01 05   Wastes from plastic shavings and turnings
12 01 06* Mineral-based machining oils containing halogens (except emulsions and solutions)
12 01 07* Mineral-based machining oils free of halogens (except emulsions and solutions)
12 01 08* Machining emulsions and solutions containing halogens
12 01 09* Machining emulsions and solutions free of halogens
12 01 10* Synthetic machining oils
12 01 12* Spent waxes and fats
12 01 13   Welding wastes
12 01 14* Machining sludges containing dangerous substances
12 01 15   Machining sludges other than those mentioned in 12 01 14
12 01 16* Waste blasting material containing dangerous substances
12 01 17   Waste blasting material other than those mentioned in 12 01 16
12 01 18* Metal sludge (grinding, honing and lapping sludge) containing oil
12 01 19* Readily biodegradable machining oil
12 01 20* Spent grinding bodies and grinding materials containing dangerous substances
12 01 21   Spent grinding bodies and grinding materials other than those mentioned in 12 01 20
12 01 99   Wastes not otherwise specified

1203 **Wastes from water and steam degreasing processes (except 11)**
12 03 01* Aqueous washing liquids
12 03 02* Steam degreasing wastes

13 **Oil wastes and wastes of liquid fuels (except edible oils and those in chapters 05, 12 nd 19)**

1301 **Waste hydraulic oils**
13 01 01* Hydraulic oils, containg PCBs
13 01 04* Oil emulsions containing organohalogenous compounds
13 01 05* Oil emulsions free of organohalogenous compounds
13 01 09* Mineral-based hydraulic oils containing organohalogenous compounds
13 01 10* Mineral-based hydraulic oils free of organohalogenous compounds
13 01 11* Synthetic hydraulic oils
13 01 12* Readily biodegradable hydraulic oils
13 01 13* Other hydraulic oils

1302 **Waste engine, gear and lubricating oils**
13 02 04* Mineral-based engine, gear and lubricating oils containing organohalogenous compounds
13 02 05* Mineral-based engine, gear and lubricating oils free of organohalogenous compounds
13 02 06* Synthetic engine, gear and lubricating oils
13 02 07* Readily biodegradable engine, gear and lubricating oils
13 02 08* Other engine, gear and lubricating oils

1303 **Waste oils and liquids used as electroinsulators and heat carriers**
13 03 01* Oils and liquids used as electroinsulators and heat carriers containing PCBs
13 03 06* Mineral-based oils and liquids used as electroinsulators and heat carriers containing organohalogenous compounds other than those mentioned in 13 03 01
13 03 07* Mineral-based oils and liquids used as electroinsulators and heat carriers free of organohalogenous compounds
13 03 08* Synthetic oils and liquids used as electroinsulators and heat carriers other than those mentioned in 13 03 01
13 03 09* Readily biodegradable oils and liquids used as electroinsulators and heat carriers
13 03 10* Other oils and liquids used as electroinsulators and heat carriers

1304 **Bilge oils**
13 04 01* Bilge oils from inland navigation
13 04 02* Bilge oils from jetty sewers
13 04 03* Bilge oils from marine navigation

1305 **Oil/water separator contents**
13 05 01* Solids from grit chambers and oil/water separators
13 05 02* Sludges from oil/water separators
13 05 03* Interceptor sludges
13 05 06* Oil from oil/water separators
13 05 07* Oily water from oil/water separators
13 05 08* Mixtures of wastes from grit chambers and oil/water separators

1307   Wastes of liquid fuels
130701* Fuel oil and diesel
130702* Petrol
130703* Other fuels (including mixtures)

1308   Oil wastes not otherwise specified
130801* Desalter sludges or emulsions
130802* Other emulsions
130880   Oily solid wastes from ships
130899* Wastes not otherwise mentioned

14   Waste organic solvents, refrigerants and propellants (except 07 and 08)
1406   Waste organic solvents, refrigerants and foam/aerosol propellants
140601* Chlorofluorocarbons, HCFC, HFC
140602* Other organohalogenous solvents and solvent mixtures
140603* Other solvents and solvent mixtures
140604* Sludges or solid wastes containing organohalogenous solvents
140605* Sludges or solid wastes containing other solvents

15   Waste packaging; sorbents, wiping cloths, filter materials and protective clothing not otherwise specified
1501   Waste packaging (including separately collected municipal packaging waste)
150101 Paper and cardboard packaging
150102 Plastic packaging
150103 Wooden packaging
150104 Metallic packaging
150105 Composite packaging
150106 Mixed packaging
150107 Glass packaging
150109 Textile packaging
150110* Packaging containing residues of or contaminated by dangerous substances (e.g. pesticides in toxicity class I and II – very toxic and toxic)
150111* Metallic packaging containing a dangerous solid porous matrix (e.g. asbestos), including empty pressure containers

1502   Sorbents, filter materials, wiping cloths and protective clothing
150202* Sorbents, filter materials (including oil filters not otherwise specified), wiping cloths (e.g. rags and dusters) and protective clothing contaminated by dangerous substances (i.e. PCB)
150203 Sorbents filter materials, wiping cloths (e.g. rags and dusters) and protective clothing other than those mentioned in 150202

16   Wastes not otherwise specified
1601   End-of-life vehicles from different means of transport (including off-road machinery) and wastes from dismantling of end-of-life vehicles, vehicle service and maintenance (except 13, 14, 1606 and 1608)
160103 End-of-life tyres
160104* End-of-life vehicles or vehicles unsuitable for use

16 01 06 End-of-life vehicles or vehicles unsuitable for use containing neither liquids nor other hazardous components
16 01 07* Oil filters
16 01 08* Components containing mercury
16 01 09* Components containing PCB
16 01 10* Explosive components (e.g. air bags)
16 01 11* Brake pads containing asbestos
16 01 12 Brake pads other than those mentioned in 16 01 11
16 01 13* Brake fluids
16 01 14* Antifreeze fluids containing dangerous substances
16 01 15 Antifreeze fluids other than those mentioned in 16 01 14
16 01 16 Tanks for liquified gas
16 01 17 Ferrous metal
16 01 18 Non-ferrous metal
16 01 19 Plastic
16 01 20 Glass
16 01 21 Hazardous components other than those mentioned in 16 01 07 to 16 01 11 and 16 01 13 and 16 01 14
16 01 22 Components not otherwise specified
16 01 99 Wastes not otherwise mentioned

16 02 Wastes from electrical and electronical equipment
16 02 09* Transformers and capacitors containing PCBs
16 02 10* Discarded equipment containing or contaminated by PCBs other than those mentioned in 16 02 09
16 02 11* Discarded equipment containing chlorofluorocarbons, HCFC, HFC
16 02 12* Discarded equipment containing free asbestos
16 02 13* Discarded equipment containing hazardous components other than those mentioned in 16 02 09 to 16 02 12
16 02 14 Discarded equipment other than those mentioned in 16 02 09 to 16 02 13
16 02 15* Hazardous components removed from discarded equipment
16 02 16 Components removed from discarded equipment other than those mentioned in 16 02 15

16 03 Off-specification batches and outdated or unsuitable for use products
16 03 03* Inorganic wastes containing dangerous substances
16 03 04 Inorganic wastes other than those mentioned in 16 03 03, 16 03 80
16 03 05* Organic wastes containing dangerous substances
16 03 06 Organic wastes other than those mentioned in 16 03 05, 16 03 80
16 03 80 Edible materials past their sell-by date or unsuitable for consumption

16 04 Waste explosives
16 04 01* Waste ammunition
16 04 02* Waste pyrotechnical materials (e.g. fireworks)
16 04 03* Other waste explosives

16 05 Gases in pressure containers and discarded chemicals
16 05 04* Gases in pressure containers (includinig halons) containing dangerous sub-stances
16 05 05 Gases in pressure containers other than those mentioned in 16 05 04

16 05 06* Laboratory and analytical chemicals (e.g. chemical reagents) containing dangerous substances including mixtures of laboratory chemicals

16 05 07* Discarded inorganic chemicals containing dangerous substances (e.g. outdated chemical reagents)

16 05 08* Discarded organic chemicals containing dangerous substances (e.g. outdated chemical reagents)

16 05 09 Discarded chemicals other than those mentioned in 16 05 06, 16 05 07 or 16 05 08

16 06 Batteries and accumulators

16 06 01* Lead batteries and accumulators

16 06 02* Ni-Cd batteries and accumulators

16 06 03* Mercury containing batteries

16 06 04 Alkaline batteries (except 16 06 03)

16 06 05 Other batteries and accumulators

16 06 06* Separately collected electrolyte from batteries and accumulators

16 07 Wastes from transport tank, storage tank and barrel cleaning (except 05 and 13)

16 07 08* Wastes containing oil or oil products

16 07 09* Wastes containing other dangerous substances

16 07 99 Wastes not otherwise specified

16 08 Spent catalysts

16 08 01 Spent catalysts containing gold, silver, rhenium, rhodium, palladium, iridium or platinum (except 16 08 07)

16 08 02* Spent catalysts containing dangerous transition metals (2) or transition metal compounds

16 08 03 Spent catalysts containing transition metals or transition metal compounds not otherwise specified

16 08 04 Spent fluid catalytic cracking catalysts used in fluidisation process (except 16 08 07)

16 08 05* Spent catalysts containing phosphoric acid

16 08 06* Spent liquids used as catalysts

16 08 07* Spent catalysts contaminated with dangerous substances

16 09 Oxidising substances

16 09 01* Permanganates (e.g. potassium permanganate)

16 09 02* Chromates (e.g. potassium chromate, sodium or potassium dichromate)

16 09 03* Peroxides e.g. hydrogen peroxide

16 09 04* Oxidising substances, not otherwise specified

16 10 Aqueous liquid wastes destined for off-site treatment

16 10 01* Aqueous liquid wastes containing dangerous substances

16 10 02 Aqueous liquid wastes other than those mentioned in 16 10 01

16 10 03* Concentrated aqueous liquid wastes (e.g. concentrates) containing dangerous substances

16 10 04 Concentrated aqueous liquid wastes (e.g. concentrates) other than those mentioned in 16 10 03

16 11 Waste linings and refractories

16 11 01* Carbon-based linings and refractories from metallurgical processes containing dangerous substances

16 11 02   Carbon-based linings and refractories from metallurgical processes other than those mentioned in 16 11 01

16 11 03* Other linings and refractories from metallurgical processes containing dangerous substances

16 11 04   Other linings and refractories from metallurgical processes other than those mentioned in 16 11 03

16 11 05* Linings and refractories from non-metallurgical processes containing dangerous substances

16 11 06   Linings and refractories from non-metallurgical processes other than those mentioned in 16 11 05

16 80   *Various wastes*
16 80 01 *Magnetic and optical information carriers*

16 81   *Wastes resulting from accidents and random incidents*
16 81 01* *Wastes displaying dangerous properties*
16 81 02 *Wastes other than those mentioned in 16 81 01*

16 82   *Wastes resulting from natural disasters*
16 82 01* *Wastes displaying dangerous properties*
16 82 02 *Wastes other than those mentioned in 16 82 01*

17   **Wastes from construction, renovation and demolision of buildings and from road infrastructure (including excavated soil and stones from contaminated sites)**

17 01   **Wastes from building materials and components and road infrastructure (e.g. cement, bricks, tiles and ceramics)**
17 01 01   Wastes of concrete and concrete debris from demolisions and renovations
17 01 02   Brick debris
17 01 03   Other wastes of ceramic materials and components of fitting
17 01 06* Mixtures of, or separate fractions of concrete, brick debris, waste ceramic materials and fitting components containing dangerous substances
17 01 07   Mixtures of concrete, brick debris, waste ceramic materials and fitting components other than those mentioned in 17 01 06
17 01 80 *Removed plasters, wallpapers, veneers, etc.*
17 01 81 *Wastes from road repairs and reconstructions*
17 01 82 *Wastes not otherwise specified*

17 02   **Wood, glass and plastic wastes**
17 02 01   Wood
17 02 02   Glass
17 02 03   Plastic
17 02 04* Wastes of glass, plastic and wood containing or contaminated with dangerous substances (e.g. railroad ties)

17 03   **Wastes of bituminous mixtures, tars and tarred products**
17 03 01* Bituminous mixtures containing tar
17 03 02   Bituminous mixtures other than those mentioned in 17 03 01
17 03 03* Tar and tarred products
17 03 80 *Waste tar paper*

17 04   **Wastes and scraps of metals and their alloys**
17 04 01   Copper, bronze, brass
17 04 02   Aluminium
17 04 03   Lead
17 04 04   Zinc
17 04 05   Iron and steel
17 04 06   Tin
17 04 07   Mixed metals
17 04 09*   Metal waste contaminated with dangerous substances
17 04 10*   Cables containing oil, tar and other dangerous substances
17 04 11   Cables other than those mentioned in 17 04 10

17 05   **Soil and stones (including excavated soil and stones from contaminated sites), and dredging spoil**
17 05 03*   Soil and stones containing dangerous substances (e.g. PCB)
17 05 04   Soil and stones other than those mentioned in 17 05 03
17 05 05*   Dredging spoil containing or contaminated with dangerous substances
17 05 06   Dredging spoil other than those mentioned in 17 05 05
17 05 07*   Track ballast (aggregate) containing dangerous substances
17 05 08   Track ballast (aggregate) other than those mentioned in 17 05 07

17 06   **Insulation materials and asbestos-containing construction materials**
17 06 01*   Insulation materials containing asbestos
17 06 03*   Other insulation materials consisting of or containing dangerous substances
17 06 04   Insulation materiale other than those mentioned in 17 06 01 and 17 06 03
17 06 05*   Construction materials containing asbestos

17 08   **Gypsum-based construction materials**
17 08 01*   Gypsum-based construction materials contaminated with dangerous substances
17 08 02   Gypsum-based construction materials other than those mentioned in 17 08 01

17 09   **Other construction, renovation and demolision wastes**
17 09 01*   Construction, renovation and demolision wastes containing mercury
17 09 02*   Construction, renovation and demolision wastes containing PCB (e.g. substances and objects containing PCB: sealants, resin-based floorings, sealed glazing units, capacitors)
17 09 03*   Other construction, renovation and demolision wastes (including mixed wastes) containing dangerous substances
17 09 04   Mixed construction, renovation and demolision wastes other than those mentioned in 17 09 01, 17 09 02 and 17 09 03

18   **Wastes from human and animal health care**
18 01   **Wastes from diagnosis, treatment and or prevention of disease in humans**
18 01 01   Instruments of surgical and other medical interventions and their remaining fragments (except 18 01 03)
18 01 02*   Body parts and organs including blood bags and blood preserves (except 18 01 03)

18 01 03* Other wastes containing live pathogenic micro-organisms or their toxins, and other forms able of transmitting genetic material that are known or suspected on credible basis to cause diseases in humans and animals (i.e. infected pampers, sanitary towels, backings) excluding 18 01 80 and 18 01 82

18 01 04 Other wastes not specified in 18 01 03.

18 01 06* Chemicals including chemical reagents containing dangerous substances

18 01 07 Chemicals including chemical reagents other than those mentioned in 18 01 06

18 01 08* Cytotoxic and cytostatic medicines

18 01 09 Medicines other than those mentioned in 18 01 08

18 01 10* Amalgam waste from dental care

18 01 80 *Spent biologically active balneotherapy baths of contagious character*

18 01 81 *Spent biologically active balneotherapy baths other than those mentioned in 18 01 80*

*18 01 82** *Wastes arising from feeding patients in isolation wards*

18 02 **Wastes from diagnosis, treatment and prevention of disease involving animals**

18 02 01 Instruments of surgical and other medical interventions and their remaining fragments (except 18 02 02)

18 02 02* Other wastes containing live pathogenic micro-organisms or their toxins, and other forms able of transmitting genetic material that are known or suspected on credible basis to cause diseases in humans and animals

18 02 03 Other wastes not specified in 18 02 02

18 02 05* Chemicals including chemical reagents containing dangerous substances

18 02 06 Chemicals including chemical reagents other than those mentioned in 18 02 05

18 02 07* Cytotoxic and cytostatic medicines

18 02 08 Medicines other than those mentioned in 18 02 07

19 **Wastes from waste management facilities, off-site waste water treatment plants and the preparation of water intended for human consumption and water for industrial use**

19 01 **Wastes from thermal processing of wastes**

19 01 02 Ferrous materials removed from bottom ash

19 01 05* Filter wastes (i.e. filter cake) from flue-gas treatment

19 01 06* Sludges and other aqueous liquid wastes from flue-gas treatment

19 01 07* Solid wastes from flue-gas treatment

19 01 10* Spent activated carbon from flue-gas treatment

19 01 11* Bottom ash and slag containing dangerous substances

19 01 12 Bottom ash and slag other than those mentioned in 19 01 11

19 01 13* Fly ash containing dangerous substances

19 01 14 Fly ash other than those mentioned in 19 01 13

19 01 15* Boiler dust containing dangerous substances

19 01 16 Boiler ash other than those mentioned in 19 01 15

19 01 17* Wastes from pyrolysis of wastes containing dangerous substances

19 01 18 Wastes from pyrolysis of wastes other than those mentioned in 19 01 17

19 01 19 Sands from fluidised beds

19 01 99 Wastes not otherwise specified

1902 **Wastes from physico/chemical treatments of wastes (including dechromatation, decyanidation, neutralisation)**

19 02 03 Premixed wastes composed only of non-hazardous wastes
19 02 04* Premixed wastes composed of at least one hazardous waste
19 02 05* Sludges from physico/chemical treatment containing dangerous substances
19 02 06 Sludges from physico/chemical treatment other than those mentioned in 19 02 05
19 02 07* Oil and concentrates from separation
19 02 08* Liquid combustible wastes containing dangerous substances
19 02 09* Solid combustible wastes containing dangerous substances
19 02 10 Combustible wastes other than those mentioned in 19 02 08 or 19 02 09
19 02 11* Other wastes containing dangerous substances
19 02 99 Wastes not otherwise specified

1903 **Stabilised/solidified wastes (3)**

19 03 04* Partly stabilised (4) hazardous wastes
19 03 05 Stabilised wastes other than those mentioned in 19 03 04
19 03 06* Solidified hazardous wastes
19 03 07 Solidified wastes other than those mentioned in 19 03 06

1904 **Vitrified waste and wastes from vitrification**

19 04 01 Vitrified waste
19 04 02* Fly ash and other flue-gas treatment wastes
19 04 03* Non-vitrified solid phase
19 04 04 Aqueous liquid wastes from vitrified waste tempering

1905 **Wastes from aerobic digestion of solid wastes (composting)**

19 05 01 Non-composted fraction of municipal and similar wastes
19 05 02 Non-composted fraction of animal and vegetable waste
19 05 03 Off-specification compost (unsuitable for use)
19 05 99 Wastes not otherwise specified

1906 **Wastes from anaerobic digestion of wastes**

19 06 03 Liquor from anaerobic digestion of municipal waste
19 06 04 Digestate from anaerobic digestion of municipal waste
19 06 05 Liquor from anaerobic digestion of animal and vegetable waste
19 06 06 Digestate from anaerobic digestion of animal and vegetable waste
19 06 99 Wastes not otherwise specified

1908 **Wastes from waste water treatment plants not otherwise specified**

19 08 01 Screenings
19 08 02 Waste from desanding
19 08 05 Stabilised municipal sewage sludges
19 08 06* Saturated or spent ion exchange resins
19 08 07* Solutions and sludges from regeneration of ion exchangers
19 08 08* Membrane system waste containing haevy metals
19 08 09 Grease and oil mixture from oil/water separation containing only edible oil and fats
19 08 10* Grease and oil mixture from oil/water separation other than those mentioned in 19 08 09

19 08 11* Sludges containing dangerous substances from biological treatment of industrial waste water

19 08 12 Sludges from biological treatment of industrial waste water other than those mentioned in 19 08 11

19 08 13* Sludges containing dangerous substances from other than biological treatment of industrial waste water

19 08 14 Sludges from other than biological treatment of industrial waste water other than those mentioned in 19 08 13

19 08 99 Wastes not otherwise specified

19 09 **Wastes from the preparation of water intended for human consumption or water for industrial use**

19 09 01 Solid waste from primary filtration and screenings

19 09 02 Sludges from water clarification

19 09 03 Sludges form decarbonation

19 09 04 Spent activated carbon

19 09 05 Saturated or spent ion exchange resins

19 09 06 Solutions and sludges from regeneration of ion exchangers

19 09 99 Wastes not otherwise specified

19 10 **Wastes from shredding of metal-containing wastes**

19 10 01 Iron and steel waste

19 10 02 Non-ferrous waste

19 10 03* Fluff-light fraction and dust containing dangerous substances

19 10 04 Fluff-light fraction and dust other than those mentioned in 19 10 03

19 10 05* Other fractions containing dangerous substances

19 10 06 Other fractions other than those mentioned in 19 10 05

19 11 **Wastes from oil regeneration**

19 11 01* Spent clay filters

19 11 02* Acid tars

19 11 03* Aqueous liquid wastes

19 11 04* Wastes from cleaning of fuel with bases

19 11 05* Sludges from on-site effluent treatment containing dangerous substances

19 11 06 Sludges from on-site effluent treatment other than those mentioned in 19 11 05

19 11 07* Wastes from flue-gas cleaning

19 11 99 Wastes not otherwise specified

19 12 **Wastes from the mechanical treatment of wastes (e.g. manual treatment, sorting, crushing, compacting, pelletising) not otherwise specified**

19 12 01 Paper and cardboard

19 12 02 Ferrous metal

19 12 03 Non-ferrous metal

19 12 04 Plastic and rubber

19 12 05 Glass

19 12 06* Wood containing dangerous substances

19 12 07 Wood other than that mentioned in 19 12 06

19 12 08 Textiles

19 12 09 Minerals (e.g. sand, stones)

19 12 10   Combustible waste (alternative fuel)

19 12 11*   Other wastes (including mixtures of substances and objects) from mechanical treatment of waste containing dangerous substances

19 12 12   Other wastes (including mixtures of materials) from mechanical treatment of waste other than those mentioned in 19 12 11

19 13   Wastes from soil and underground water remediation

19 13 01*   Solid wastes from soil remediation containing dangerous substances

19 13 02   Solid wastes from soil remediation other than those mentioned in 19 13 01

19 13 03*   Sludges from soil remediation containing dangerous substances

19 13 04   Sludges from soil remediation other than those mentioned in 19 13 03

19 13 05*   Sludges from underground water remediation containing dangerous substances

19 13 06   Sludges from underground water remediation other than those mentioned in 19 13 05

19 13 07*   Aqueous liquid wastes and concentrated aqueous liquid wastes (i.e. concentrates) from underground water remediation containing dangerous substances

19 13 08   Aqueous liquid wastes and concentrated aqueous liquid wastes (i.e. concentrates) from underground water remediation other than those mentioned in 19 13 07

19 80   Wastes from neutralisation of medical and veterinary wastes not otherwise specified

19 80 01   Wastes from autoclave treatment of medical and veterinary wastes

20   Municipal wastes including separately collected fractions

20 01   Municipal wastes sorted and collected separately (except 15 01)

20 01 01   Paper and cardboard

20 01 02   Glass

20 01 08   Biodegradable kitchen waste

20 01 10   Clothes

20 01 11   Textiles

20 01 13*   Solvents

20 01 14*   Acids

20 01 15*   Alkalines

20 01 17*   Photographic reagents

20 01 19*   Pesticides in toxicity class I and II (very toxic and toxic, i.e. herbicides, insecticides)

20 01 21*   Fluorescent tubes and other mercury-containing waste

20 01 23*   Discarded equipment containing chlorofluorocarbons

20 01 25   Edible oil and fat

20 01 26*   Edible oil and fat other than those mentioned in 20 01 25

20 01 27*   Paint, drawing and printing inks, sealants, adhesives and resins containing dangerous substances

20 01 28   Paint, drawing and printing inks, sealants, adhesives and resins other than those mentioned in 20 01 27

20 01 29*   Detergents containing dangerous substances

20 01 30   Detergents other than those mentioned in 20 01 29

20 01 31*   Cytotoxic and cytostatic medicines

20 01 32   Medicines other than those mentioned in 20 01 31

200133* Batteries and accumulators included in 160601, 160602 or 160603 and unsorted batteries and accumulators containing these batteries
200134   Batteries and accumulators other than those mentioned in 200133
200135* Discarded electrical and electronical equipment other than those mentioned in 200121 and 200123 containing hazardous components (1)
200136   Discarded electrical and electronical equipment other than those mentioned in 200121, 200123 and 200135
200137* Wood containing dangerous substances
200138   Wood other than that mentioned in 200137
200139   Plastics
200140   Metals
200141   Waste ventillation sweepers
200180   Pesticides other than those mentioned in 200119
200199   Other fractions collected separately not otherwise specified

2002   **Wastes form gardens and parks (including cemetery waste)**
200201   Biodegradable waste
200202   Soil and stones
200203   Other non-biodegradable wastes

2003   **Other municipal wastes**
200301   Unsorted (mixed) municipal waste
200302   Waste from markets
200303   Street and square cleaning residues
200304   Sludge from tanks without drainage for collecting sewage
200306   Waste from interceptor sewers
200307   Bulky waste
200399   Municipal wastes not otherwise specified

Used annotations

[0] First two digits define a group of wastes identifying the source generating the waste. The definition of a group of wastes together with next two digits identifies a subgroup of wastes, and the code consisting of six digits identifies a specific waste. Any waste marked with an asterisk (*) is considered as a hazardous waste pursuant to separate regulations on hazardous wastes.

[1] Hazardous components from electrical and electronic equipment may include accumulators and batteries mentioned in 1606 and marked as hazardous; mercury switches, glass from cathode ray tubes and other activated glass, etc.

[2] Transition metals include: scandium, vanadium, manganese, cobalt, copper, yttrium, niobium, hafnium, tungsten, titanium, chromium, iron, nickel, zinc, zirconium, molybdenum and tantalum. These metals and their compunds are dangerous if they are classified as dangerous substances. The classification of dangerous substances is laid down in separate regulations.

[3] Stabilisation processes change the dangerousness of the constituents in the waste and thus transform hazardous waste into non-hazardous waste. Solidification processes only change the physical state of the waste (e.g. liquid into solid) without changing the chemical properties of the waste.

(4) A waste is considered as party stabilised if, after the stabilisation process, dangerous constituents which have not been changed completely into non-dangerous constituents could be released into the environment in the short, middle and long term.

Comments to Appendix 4 in English edition

The European Community directives on wastes and hazardous wastes have been implemented in the Polish legislation with subsequent amendments and adopted as legally binding acts by the Lower House of the Polish parliament, Senate and the Council of Ministries. These amendments introduce a number of waste subgroups that were originally not included in the list of wastes and hazardous wastes established by the Commission of the European Communities. In order to discern these additional subgroups of wastes from those originally adopted by the Commission the corresponding subchapters of the Polish catalogue of wastes were written in *italics*. In several cases there are also some differences between the wording used in the European Community list of wastes and the Polish version of this catalogue. In general the introduced words and phrases are equivalent to the original wording or even refine the meaning of particular waste categories. There are, however, some statements that are not quite consistent with the original ones or display unjustified omittments. Below the most significant discrepancies are listed in detail:

02 01 06   originally specified list of animal wastes is abbreviated to only one item

02 03   **conserve production is left out**
02 04 01   sludges instead of soil
03 02 99   wastes instead of wood preservatives
03 03 02   originally only green liquor sludge from recovery of cooking liquor
04 01 08   blue sheetings are omitted
04 02 19   wastes instead of sludges
04 02 20   same as above

06   **industry instead of processes**
06 07 04   sulfuric instead of contact
06 08 02   chlorosilanes replaced by silicones. Silanes are compounds with general formula Si_nH_{2n} which are known as branched or unbranched (up to n = 8). Substitution of at least one hydrogen atom by chlorine atom gives an appropriate chlorosilane. Hydrolysis of some di- and trichloro-dimethylsilanes may lead to branched or bridged siloxanes. Silicones are polymers produced commercially from siloxanes. They can be made as fluids (oils), greases, emulsions, elastomers (rubbers), and resins. In this chain of events chlorosilanes are substrates, siloxanes are intermediates, and silicones are products.

07   **industry instead of processes**

07 07   **fine chemicals are left unmentioned**
09 01 06   on-site effluent treatment instead of on-site treatment. Such requirement with respect to small workshops providing photographic services seems to be an exaggeration.
09 01 08   silver compounds are omitted

10 01 18	flue-gas treatment instead of gas cleaning
10 01 19	same as above
10 02 01	just a slag from smelting but not the wastes from the slag processing
10 02 07	flue-gas instead of gas. It implies that only this specific gas undergoes such treatment.
10 02 08	same as in 10 02 07
10 02 13	same as in 10 02 07
10 02 14	same as in 10 02 07

10 03   **adjective thermal omitted, thus the subchapter refers both to hot and cold metallurgy**

10 03 23	same as in 10 02 07
10 03 24	same as in 10 02 07
10 03 25	same as in 10 02 07
10 03 26	same as in 10 02 07

10 04   **same as in 10 03**

10 04 02	dross is left out
10 04 06	same as in 10 02 07
10 04 07	same as in 10 02 07

10 05   **same as in 10 03**

10 05 05	same as in 10 02 07
10 05 06	same as in 10 02 07
10 05 10	same as in 10 04 02
10 05 11	same as in 10 04 02

10 06   **same as in 10 03**

10 06 02	same as in 10 04 02
10 06 06	same as in 10 02 07
10 06 07	same as in 10 02 07

10 07   **same as in 10 03**

10 07 02	same as in 10 04 02
10 07 03	same as in 10 02 07
10 07 05	same as in 10 02 07

10 08   **same as in 10 03**

10 08 10	same as in 10 04 02
10 08 11	same as in 10 04 02
10 11 11	electron instead of cathode. This is understandable since cathode ray tubes are now commonly replaced by electron ray tubes.
10 12 09	same as in 10 02 07
10 12 10	same as in 10 02 07

10 13   **articles are omitted**

10 13 07	same as in 10 02 07
10 13 12	same as in 10 02 07
10 13 13	same as in 10 02 07
10 14 01	same as in 10 02 07

11 01   **etching as similar to pickling is omitted**

11 05 03   same as in 10 02 07
13 01 04   chlorinated emulsions are replaced by oil emulsions containing organo-ha-
            logenous compounds. This entry expands the group of originally considered
            compounds.
13 01 05   similar to 13 01 04
13 01 09   similar to 13 01 04
13 01 10   similar to 13 01 04
13 02 04   similar to 13 01 04
13 02 05   similar to 13 01 04
13 03 06   similar to 13 01 04
13 03 07   similar to 13 01 04
13 04 03   marine instead of other

16 03   **unused replaced by outdated and unsuitable for use**

17 03   **tars instead of coal tar**
17 03 01   same generalization as in 17 03
17 03 03   same generalization as in 17 03
17 04 10   same generalization as in 17 03

18 01   **natal care left out**
18 01 01   original short expression from medical jargon 'sharps' replaced by much lon-
            ger description but missing the point. Sharps include needles, syringes, and
            lancets mostly to manage medical conditions at home. The common feature
            of all these objects is their sharpness which is not emphasized in the Polish
            translation. The used sharps such as for instance needles or broken glassy am-
            poules display dangerous properties. People exposed to sharps face not only
            the risk of a painful stick but also the risk of contracting a life-alternating
            disease such as HIV/AIDS or Hepatis B or C.

18 02   **for unknown reasons the originally introduced research is here not taken
            into account.**
18 02 01   same as in 18 01 01

19 01   **incineration or pyrolysis of waste is replaced by thermal processing of
            wastes which is a more general phrase**
19 01 05   same as in 10 02 07
19 01 06   same as in 10 02 07
19 01 07   same as in 10 02 07

19 05   **digestion instead of treatment**

19 06   **same as in 19 05**
19 06 03   same as in 19 05
19 06 04   same as in 19 05
19 06 05   same as in 19 05
19 06 06   same as in 19 05

19 07   **whole subchapter (on landfill waste) is omitted in the Polish implementa-
            tion**
19 08 05   sludges from treatment of urban waste water replaced by stabilised municipal
            sewage sludges.

19 11 01   spent filter clays changed to spent clay filters. Erroneous transposition of two neighboring words in the Polish translation.

19 13   underground water in place of ground water. During remediation of the soil the remediation of the water that is closer to the surface seems to be more important.

19 13 05   same as in 19 13
19 13 06   same as in 19 13
19 13 07   same as in 19 13
19 13 08   same as in 19 13

20   specified forms of municipal wastes omitted
20 01 08   only kitchen left from kitchen and canteen
20 01 41   wastes from chimney sweeping replaced by waste ventilation sweepers. This looks like non-adequate translation. Contrary to the soot from the chimney the left fragments of cleaning tools constitute just a less significant waste.
20 03 06   interceptor sewers instead of sewage cleaning.

Throughout the whole catalogue of wastes absorbents are replaced by sorbents. This is an extension of the original term since sorbents include absorbents and adsorbents as well.

9.5. Appendix 5 – Waste Transfer Card

WASTE TRANSFER CARD		Card No.[a]	Year by the calendar
Waste holder passing the waste on[b,c]	Transporter performing activities in the waste transportation[b, d]		Waste transfer facility owner receiving the waste[b]
Address[e]	Address[d,e]		Adresse[e]
Phone/fax	Phone/fax[d]		Phone/fax
REGON No.	REGON No.[d]		REGON No.
Waste location address[f]			
Waste code	Waste type		
Day/month[g]	Weight of transferred waste [kg][h]		Registration number of vehicle, trailer or wagon[d,i]
Endorsed by waste holder passing the waste on	Endorsed by transport operator servicing the waste transfer[d]		Endorsed by waste recipient
Date, signature, stamp	Date, signature, stamp		Date, signature, stamp

Legend

[a] Number given by the waste holder passing the waste on.

[b] Person's full name or company name.

[c] In case of municipal wastes the card is filled out by an entrepreneur who received a permit from the owners of fixed properties for conducting the waste management activities as specified in the act of 13th September 1996 on maintaining cleanness and order in municipalities (JL of 2005 No. 236, item 2008), or by a communal subunit as specified in the Act of 13th September 1996 on maintaining cleanness and order in municipalities.

[d] In case when the waste is transported subsequently by two or more transporters conducting the waste transportation activities the requested data and signatures of all the transporters involved in the waste transportation should be collected in the provided spaces retaining the proper waste transportation sequence.

[e] Person's or company location address.

[f] Location address of the waste reception indicated by the waste holder to the transporter conducting the waste transportation activities where the waste should be delivered.

[g] In case of hazardous wastes enter the date of the waste transfer. The card may be used as a one-time waste transfer card or as an overall waste transfer card including the same waste transferred during one calendar month by the same transporter to the same waste transfer facility owner.

[h] Enter the weight of the wastes with accuracy of at least one significant number for the non-hazardous wastes and at least three significant numbers for the hazardous wastes.

[i] Applies to hazardous wastes.